Adventures in Homesteading

Black's Tropical Homestead

Bobby and Sherrie Black

Cushing Publishing
www.cushingpublishing.com

Copyright © 2024 Black's Tropical Homestead
ISBN: 978-1-963661-12-5

Cushing Publishing
9380 Driver Road
Middlesex, NC 27557

All rights reserved. No part of this book may be reproduced, stored, or transmitted by any means—whether electronic, mechanical, auditory, written, or graphic—without the written permission of both the author and publisher, except for excerpts required for reviews and articles. Unauthorized reproduction of any part of this work is illegal and punishable by law.

Dedicated to our loving Mothers, Shirley Black and Mary Jamison, who planted the seed of knowledge in our minds and nurtured it.

To those rare, beautiful friendships that change our lives forever, Louisa Abbot, Knesha Calvin, Jerry Gibbons, and Mark Scott.

To the ones reading this, may you find the courage, joy, and love contained within these pages.

Last but not least, in the loving memory of our fathers Bobby Black Sr and Rudolph Jamison I, whose stories live on in these pages forever and always.

Table of Contents

What's Up, Soil Brothas and Soil Sistas?................7
How Did All This Start?........................12
The Pandemic......................14
What Did We Do First?........................17
South Florida..........................20
Chicken Math Is Real....................23
Benefits of Our Apothecary....................29
Homesteading Housewife......................40
Favorite Plants.........................42
Hurricane Season....................44
Social Media, Really?........................45
Community, Yeah....................47
Raised the Beds......................48
Rotten Rottweiler.....................49
Let's Do the Bunny Hop....................51
Kitchen Scientist....................58
Homesteading........................60
SoilFamily Expo......................62
To Bee or Not to Bee...................67
This Little Piggy....................70
SoilFamily Wow.....................75
Freeze Drying Fun....................78
The Summer of Help....................80
Our Grocery Store....................84
Grow What You Eat....................86
So Fresh So Clean....................93
Home Pantry......................95
Oreo Cookies......................96
Cinnamon Roll......................98
Vanilla Ice Cream....................100
Sandwich Bread....................101
Peanut Butter Cup....................103

Honey Caramel..106
Chocolate..107
Ginger Bug...108
Shampoo..111
Hair Conditioner..112
Apple Cider Vinegar..113
Laundry Soap...114
Dish Detergent...115
Fire Cider..116
Elderberry Syrup...117
Focus Tea Blend..119
Kombucha...120
Lemon Oreo Pound Cake..................................124
Natural Immune Booster..................................127
Watermelon Jam..128
Frosty...129
Homemade Cheese...130
Root Beer Soda...132
Bath Cleaning Bombs..134
Biography..135

What's Up, Soil Brothas and Soil Sistas?

Bobby

I am Bobby Black, and my beautiful wife is Sherrie Black. We have some amazing and beautiful things we would love to share with everyone. Let's start by telling all y'all beautiful people a little about the Blacks. We got our social media name BLACK'S TROPICAL HOMESTEAD from our last name and our special love for growing.

We raise our meat, grow a lot of our own vegetables, and make homemade items to be as self-sufficient as possible. This is the beautiful and amazing thing we want the world to know about, to help you understand that you too can live this lifestyle right where you are. We hope to spark a fire in someone who's on the fence to get motivated.

Homesteading can provide healthy food to eat, improve mental health, bring your family closer together, and open the door to a community of beautiful people. You will be creating the environment to create life. You will be nurturing both plants and animals, giving them a really good life, then harvesting them and feeding your body and those of your family and friends. Doing so will give you a greater respect for your food and your life, giving it a whole new meaning.

If you are meeting us for the first time, you might think we were raised in the country and our families taught us how to do all these things. However, we were not raised with any of the knowledge or skills we will talk about. We were both city kids. I, Bobby, your

ADVENTURES IN HOMESTEADING

Soil Brotha, am from Savannah, Georgia. My mother started life in the country and my father was from the city. Sherrie, your Soil Sista, is from Virginia, the youngest of six children and the only girl. Her mother and father were born and raised in the city. While I'm sure the knowledge of natural living was known to our parents, I think theirs was the generation that stopped passing these things down to their kids because they didn't feel they were necessary anymore. We both are still in the city for now and may always be.

Then the COVID-19 pandemic hit, and it made us reevaluate our lives and the way we live them. We felt the food recalls and the deaths and illnesses that stemmed from them were cause for concern. The number of preservatives and chemicals we consume daily should make us wonder what they are doing to our long-term health, both mentally and physically.

While it's almost impossible for most of us to avoid many of these preservatives and chemicals because they are in everything, we can do better. We want to share our story with everyone, and hopefully we can inspire anyone who is unsure that if we can do this, you can too. Your journey may look different from ours, and that's okay. The path to getting where you want to be must start from where you currently stand.

It is funny that a lot of people think our name stems from the fact that we are African American. We get that question a lot and I'm sure it makes some hesitant. People assume that what we do is just for a particular race. But this is not the case.

What we do is for a particular type person, a peaceful person, a loving person, a considerate person, a person who wants to look at the positive side of things even though it's not always easy, a person who would like to build up and not tear down others, people that

BOBBY AND SHERRIE BLACK

respect differences, eat the meat, spit out the bones. No one race has a monopoly on any of this.

Your best help, your most important lessons, your greatest support, and the ah-ha moments can come from any person. If you exclude people just because they don't look like you, you are missing the opportunity to be the best you can be. What positive impact do you have on the lives of others? What impact do you have on people when you are not there? Here's a big one: will people have to lie about how good a person you were after you have passed? Instant gratification will usually only be given with money. All the rest, you just have to live it and have faith that you left a positive impact on others' lives.

One of our greatest pleasures is receiving comments about families that have started growing their own gardens. Parents will tell us that they bring their kids to the garden with them and teach them how to nurture it. It makes us feel so good to know that these kids will not grow up not knowing anything about growing their own food.

The amazing thing is these kids can grow up to be anything, from our next farmers to people in power who

ADVENTURES IN HOMESTEADING

may make important decisions in our country or the world about our foods. And don't we want the people in control of our food to know what they're doing? We can't make good decisions about things of which we are ignorant, no matter how many people try. And it's possible that we, in sharing our enthusiasm, and these amazing parents, by including their children in their gardens, can create such an outcome.

The world does everything it can to turn all of us against one another by putting us into categories: political parties, sexual orientation, race, religion, financial brackets. Why focus on the one thing that makes us different and ignore the many ways in which we are the same?

That one thing may mean we can't be close friends, but it doesn't mean that someone is evil. Don't get me wrong, there are some very nasty people in this world, and you might be one of them, but they're not the majority. You will never see us searching for a reason to not identify with someone. We do notice differences when they appear, but if they're not dangerous, let's try to live and let live. Some ask us what Soil Brothas, Soil Sistas, and Soil family are. Sherrie, your Soil Sista, came up with these terms, and when she told me, I thought they were perfect.

I saw the word Soil as a metaphor for the foundation for growth in all ways. We consider you Soil family if you are doing that work to grow your food, relationships, financial status, your faith, and yourself. With the perfect soil—or soil being worked on—the possibility for growth is endless. This became a term of endearment to those in our community.

BOBBY AND SHERRIE BLACK

How Did All This Start?

Bobby

So, without getting all in the weeds, I have been growing food for a little over twenty-five years, mainly small gardens, and banana plants. I've always found great joy in doing this. I'm a peaceful person and am always in search of serenity. Looking for peace and peaceful people was my way of controlling how I navigated this world. Gardening was one of the things that helped bring me peace, and no matter how hard life worked to disrupt it, I found ways—working out, fishing, and of course gardening—to help reduce the stress in my everyday life. Before the pandemic, I had a small garden growing on the side of the house, and I had about eight different varieties of banana plants.

About twenty-five years ago, I went to a local bamboo farm in Savannah where the farmers were doing some research for the University of Georgia. They were trying to see if bananas could grow as a crop in the Georgia climate. When I arrived, my attention was immediately caught by a field full of beautiful plants with huge leaves. There was a gentleman

BOBBY AND SHERRIE BLACK

out there who explained to me that these were banana plants.

Of course, I could not wrap my mind around what he had said because I had no idea how bananas grew and what the plants looked like. After I asked a million questions and saw bananas hanging from the plants, I was hooked. I returned just to look at them and walk under them about twice a week for several years. I learned as much as could about them and then I started growing them myself. They were also growing dragon fruit, and after a few years, I finally felt ready to try growing dragon fruit.as well, so I ordered some dragon fruit cuttings. Sherrie would come out occasionally to look for a few minutes, but that was about it. She couldn't get over the bugs, ha!

The Pandemic

Bobby

In 2021, we kept hearing talks of shutdowns due to the COVID-19 pandemic. The week before lockdown, we decided to bring an elderly friend home from the nursing home who Sherrie was helping take care of before the pandemic. We wanted to make sure she was being properly looked after, and she was Sherrie's sole focus during the time she lived with us.

During this time, we had an opportunity to purchase our little girl Nala, a female Rottweiler. Rottweilers are my favorite breed, and I've had this breed for thirty years. I love their intellect, their usefulness as protectors, and how loving, trainable, and affectionate they are. Rottweilers can be teddy bears that can turn into beasts if they need to. Sherrie also loves the breed and had been around Rottweilers before.

We enjoyed staying home raising our little girl, taking care of our friend, and enjoying each other. We seemed to be one of the few couples that thrived during the pandemic. We can be around each other all the time and enjoy it. Sherrie and I have a unique dynamic. I noticed shortly after we met that my strengths were her weaknesses and vice versa. After I discovered that she was the kind of woman I was looking for, we were inseparable. Together we can do anything.

Our friend left us at 108 years old. Ms. West was a beloved part of our family, and we were very sad to lose her. She lived a very long healthy and happy life and passed peacefully with us by her side. She

BOBBY AND SHERRIE BLACK

lived most of her life on an island where a lot of her food was grown and raised, and she had no major health issues. This made us start considering how beneficial such a lifestyle could be.

Sherrie

After losing Lovie [Ms. West], Bobby did not want me returning to the medical field working in hospitals and nursing homes after the pandemic restrictions were lifted. Plus, I was dealing with a major back injury. It was just too dangerous, so we decided for me to become a full-time housewife. No stress from the outside world and moving at my own pace, I was perfectly fine with that. So, I began to learn more about gardening by getting out in the yard with Bobby.

Me outside, gardening! Ha, I still laugh at the thought. I grew up in a small city in Virginia. I don't remember my parents having a garden. I do remember always being afraid of bugs, frogs, and snakes. I don't even recall seeing lizards growing up. And of course, living in the south, we had to deal with gnats and mosquitoes. They always want to drink my blood. Not Bobbie's just mine, ha!

At the house, my time outside was limited. I was afraid of everything moving except a human. Being afraid of bugs and animals is an adventure in itself. I had to depend on my husband to save me from the

ADVENTURES IN HOMESTEADING

frogs, snakes, lizards, and many other things.

Bobby would always say, "It's okay, they are here to help!" But have you ever seen the giant water bugs here in Georgia? They are terrifyingly disgusting. But over time I've gotten used to the bugs. I stay out of their way and pray they stay out of mine. Teamwork!

I went from a forty-three-year-old woman who only ever owned a dog to one who now owns dogs, fish, chickens, rabbits, quail, and pigs, and I absolutely love it. I told Bobby a miniature cow is in my near future! I love animals and I would love to one day have a farm full of them.

I had a lot to learn and tons of fears to overcome. I was terrified of the bugs, not to mention snakes, frogs, lizards, and worms. And let's not forget the mosquitos love me to death! Between digging holes and the bugs, I'd disappear fast! I'd leave Bobby outside and go sit in the bathroom for an hour or two! When I returned, he'd laugh and ask where I was. I'd tell him I was in the bathroom—I was a little constipated. Ha!

What Did We Do First?

Bobby

Sherrie began asking questions about my dragon fruit cuttings as I was waiting for the weather to warm up so they could start growing. I explained to her all the different varieties and the different tastes. We did our research and began the journey of growing dragon fruit.

We got large pots and drilled drain holes in the bottoms. Then we got a five-foot-long three-inch PVC pipe. We then got a one-inch pipe and cut them to the length of the bottom of the pot. We got a pig panel that we cut into a manageable piece. We cut slots in the bottom of the PVC pipe the size of the one-inch pipe. We slid the one-inch pipe inside going in four different directions. We then put the five-inch pipe up in the pot. We poured a layer of river rock into the pot to hold the PVC pipe sturdy and porous enough for good drainage.

We then placed the pig panel on the top. We then filled the pots with good potting soil. We put the cutting up in the pot to root. Just like that, our first two dragon fruit plants were planted.

I was amazed that Sherrie was there every step of the way, as before I'd been the only one doing this, and it felt good to have someone to share it with. I had someone who wanted to grow with me and was excited to do so. I'd never had this experience.

Like most people, she had never heard of all these tropical fruits before doing the research. When Sherrie

ADVENTURES IN HOMESTEADING

gets interested in something, I either move it out of her path, pull her back, or get out of her way. And since this was something, I wanted to share with her, I got out of her way.

Our plants grew great and even flowered. We were setting alarms to get up at night so we could hand-pollinate our nighttime-blooming dragon fruit. We'd assumed that the plants we had were self-fertile, but we found out the hard way they were not. One plant flowered over twenty times while the other never flowered, so we lost all our fruit.

Before I knew it, we were in all kinds of tropical growing groups. Sherrie was talking regularly with growers from our native Georgia to Florida to California. We were receiving cuttings of different varieties of dragon fruit and other unusual tropical plants from all over.

We began driving for many hours to visit people we met online who were growing dragon fruit. I think we ended up with twenty-three different varieties, not to mention so many other tropical plants. If we were not together, I'm confident that this would not have happened.

We also started growing a lot of passion fruit at the time. We made trellises out of T-posts and pig panels and planted them two at a time on the structures. Sherrie and I were able to eat this amazing fruit that we'd helped to grow, and we fell in love with their taste. When we found that they go for five or six dollars in the store we were amazed. I could eat these off the tree, and Sherrie preferred them on the cake, in ice cream, or made into sweet juice.

We are now harvesting buckets of passion fruit. This was also the first time Sherrie ever saw bananas grown and harvested, and she was just as in love with the process as I was. Sherrie was not aware of the different varieties and flavors that bananas had. There was no stopping her now.

South Florida

Bobby

My mind is always working! I ask myself, "What can I do to provide my family and friends with a healthy life?" Watching Lovie grow older and pass away so peacefully on her 108th birthday, I knew we had to make a lifestyle change! We never know what God has in store for us. Is it 108-plus years of living? We just don't know, but what I do know is we have to take care of ourselves now by living mainly off the land! And that's exactly what we're doing on our beautiful one-acre suburban homestead!

I first started researching the plants we were growing like miracle berry, soursop, mango, papaya, shampoo ginger, turmeric, and other tropical fruits, and I fell in love with my findings! Most of these plants offer some sort of medicinal properties that we or our family, friends, and neighbors may need one day.

Sherrie

We traveled to South Florida and found so many unique and beautiful plants that are also beneficial, not just for the fruit but also for the leaves, stems, and roots, and I needed them all! Bobby would always say I was doing too much at

BOBBY AND SHERRIE BLACK

one time, but I'd remind him, "We need these plants, babe, and we may not come back to South Florida for a year and may never find these plants again."

We'd walk the nurseries and research different plants to be sure we knew what we were purchasing and why. We'd shop until the car was full of plants, leaving just enough space for Bobby and me to fit in the car and make the seven-hour trip back to Savannah. So our apothecary grew fast.

Learning what part of the plant we can make different remedies with just made me want to learn more. From tinctures, tea blends, salves, and so much more, I'm ready to learn as much as my brain can store.

ADVENTURES IN HOMESTEADING

Chicken Math Is Real

Sherrie

Chickens are so cute, especially as babies. However, I'd never even touched a chicken. I thought to myself that they might eat me! Every time we'd go to a feed store, I'd admire the baby chicks. Finally, one day in Tractor Supply, Bobby went to the restroom and I stood there checking out those precious babies. My mind was set! Bobby returned, and I said, "Let's get chickens!" I was a little worried because yes, I was scared of them, but they were babies and I'd get over this fear.

Bobby said, "You can have chickens if you are going to take care of them." Ha! Okay, let's do this. Now in my mind, I was thinking he'd take care of them, but nope, it was my job.

ADVENTURES IN HOMESTEADING

Bobby

Up to that point, I haven't even thought about chickens. I initially said no. Sherrie asked again and I said, "If we get chickens, you're the one who will take care of them." She agreed. We did our research to find the best chickens for our area. I thought at that time I wouldn't mind having fresh eggs, but I didn't even think how much work I was going to have to do to get ready to have chickens.

So, we purchased two Rhode Island Reds, two Isa Browns, and two Leg Horns, and that was it. Or so I thought. Nope: Sherrie came home with three more. Surely that would be it. Nope: now she and I realized that chickens lay green, blue, and red eggs, and in all different shades of those colors. I don't know why we had to order twelve Easter Eggers. Long story short, we were supposed to get four chickens and ended up with twenty-one.

I didn't see anything for sale that would hold this number of chicks. I'm no carpenter and had few carpentry tools. I went online and was able to decipher some plans for a chicken coop. Meanwhile, chickens were starting to hop out of the makeshift brooder box inside the house. We weren't prepared for this.

We then purchased the wood needed to build coop, and due to pandemic shortages, wood was at the price of gold at the time. We built a four-foot by four-foot coop three feet off the ground. We purchased a six-foot dog kennel and attached it to the chicken coop. Then we wrapped it all with hardware cloth. For added protection, we laid chicken wire on the ground all around the coop, and from there, we ran with it.

This was an initial investment that I wasn't expecting.

BOBBY AND SHERRIE BLACK

We enjoyed the girls, who produced a lot of colorful, beautiful eggs. We weren't expecting to have as many eggs as we got; we had up to twenty dozen at one time. I eat a lot of eggs, but I couldn't eat this many! Thankfully, we had a few people to take some.

Without realizing it, we had fallen into a pattern of spending every moment together, walking the yard, raising the chickens, and tending to the garden and plants. I told Sherrie that we should get meat birds so we could raise our chickens to be eaten. I felt it would not be an issue, being a hunter, to process food to eat. I remember seeing a look of surprise on her face, and she said she didn't think she could eat them. Even so, we ordered twenty Cornish Cross chicks. I built a chicken tractor to hold them, and we fed them well. As the time came closer, I decided to purchase a chicken plucker to make the process easier. I built some cones from instructions I found online.

When they were ten weeks old, it was time to process. I set up different stations to make processing easier. We had the cone station where the chickens slowly passed away. We then removed from the cone station and dipped them into hot water (176 degrees) for about twenty to thirty seconds.

After removing the chicken from the hot water, we dropped them in the chicken plucker. We sprayed the chicken in the machine while all the feathers were removed. The feathers were rejected from the bottom of the machine. In thirty to forty-five seconds, we removed our bird, naked. The chicken plucker isn't mandatory, but it made the process so much faster and easier.

Now it was time to go to the processing table. I put the heart, liver, and gizzards in a bowl of ice water. We used the innards, the feathers, and other parts for

ADVENTURES IN HOMESTEADING

dog treats or fertilizer. To my surprise, Sherie was out there helping me, and we got the chicken processed in no time. I knew I had one chance to convince her that this was worth doing by preparing a good meal. We froze all the chickens. A couple of days later the heat was on. I decided I was going to rotisserie the chicken. Using an injector, I seasoned it to my liking. It smelled so good, and I just knew this was going to be amazing. I broke down the chicken after cooking, and it was very tender. I had a side of mashed potatoes and green beans. It tasted so amazing that I forgot to see how Sherrie was doing with hers. When I looked, there were just bones on her plate. She said "that it was the best chicken she'd ever had," and the rest is history. We ate the whole chicken that night.

Now we are growing and harvesting fruit, vegetables, and meat, and we plan to continue doing this on the regular. With the addition of our chickens, we became homesteaders. We may not have had a lot of land, but our minds and actions were that of homesteaders.

Without realizing it, we had fallen into a pattern of spending every moment together, walking the yard, raising the chickens, and tending to the garden and plants. I told Sherrie that we should get meat birds so we could raise our chickens to be eaten. I felt it would not be an issue, being a hunter, to process food to eat. I remember seeing a look of surprise on her face, and she said she didn't think she could eat them. Even so, we ordered twenty Cornish Cross chicks. I built a chicken tractor to hold them, and we fed them well. As the time came closer, I decided to purchase a chicken plucker to make the process easier. I built some cones from instructions I found online.

When they were ten weeks old, it was time to process. I set up different stations to make processing easier. We had the cone station where the chickens slowly

BOBBY AND SHERRIE BLACK

passed away. We then removed from the cone station and dipped them into hot water (176 degrees) for about twenty to thirty seconds.

After removing the chicken from the hot water, we dropped them in the chicken plucker. We sprayed the chicken in the machine while all the feathers were removed. The feathers were rejected from the bottom of the machine. In thirty to forty-five seconds, we removed our bird, naked. The chicken plucker isn't mandatory, but it made the process so much faster and easier.

Now it was time to go to the processing table. I put the heart, liver, and gizzards in a bowl of ice water. We used the innards, the feathers, and other parts for dog treats or fertilizer. To my surprise, Sherie was out there helping me, and we got the chicken processed in no time. I knew I had one chance to convince her that this was worth doing by preparing a good meal. We froze all the chickens. A couple of days later the heat was on. I decided I was going to rotisserie the chicken. Using an injector, I seasoned it to my liking. It smelled so good, and I just knew this was going to be amazing. I broke down the chicken after cooking, and it was very tender. I had a side of mashed potatoes and green beans. It tasted so amazing that I forgot to see how Sherrie was doing with hers. When I looked, there were just bones on her plate. She said "that it was the best chicken she'd ever had," and the rest is history. We ate the whole chicken that night.

Now we are growing and harvesting fruit, vegetables, and meat, and we plan to continue doing this on the regular. With the addition of our chickens, we became homesteaders. We may not have had a lot of land, but our minds and actions were that of homesteaders.

ADVENTURES IN HOMESTEADING

Benefits of Our Apothecary

Sherrie

Herbs are food and medicine! We've done so much research about different herbs, and I must admit we were both shocked to find out most synthetics contain some type of herb along with their other ingredients. God has provided us with everything we need to live and be healthy. We just have to find the information on the proper way to use it. Let's discuss a few herbs, their benefits, and a few ways to use them.

We suggest you do your own research on these herbs as we are not doctors and cannot give medical advice.

Cayenne pepper is one of my favorite herbs. We take a cayenne tincture daily and we keep cayenne salve around for muscle aches and pains. Cayenne is said to help boost immunity, prevent nasal congestion, help aid in digestion, lower asthma risk, enhance blood circulation, tackle arthritis, lower blood pressure, aid in weight loss, relieve toothache, relieve pain, reduce inflammation, and so much more.

Cayenne pepper or tincture can also be added to teas and soups and be used to season foods. I drop a few cayenne peppers into a jar of honey to make Sweet Heat. It's the best drizzle on fried chicken, pork chops, or rabbit. Cayenne pepper has been known to save lives: it can help prevent or stop a heart attack!

For years, I was annoyed with our pine trees. Then I found a few ways I could use the needles. For one, I make our pine cleaner with pine needles and distilled water vinegar. Simply cram a jar full of pine needles

ADVENTURES IN HOMESTEADING

and fill it with vinegar. I like to cut up a lemon also and throw it in the jar. I let it set for a few weeks then strain and use it. I dilute mine with equal parts water, and just like that, you have a safe and very effective cleaning solution.

Pine tea is very beneficial for the mind and body also. Pine offers vitamin A to the body, help improve eyesight, help beautify the skin and hair, improve red blood cells, reduce inflammation, relieve arthritis aches, pains, and soreness, and help with mental clarity.

Moringa! The tree of life. A daily dose of moringa can definitely change your life. We freeze dry enough moringa every year to help get through the summer. Moringa can be used in seasoning foods, soup, tea, and smoothies. I like to eat the leaves fresh or in my salad.

Moringa has a wide range of nutritional and bioactive compounds, including essential amino acids, carbohydrates, fiber, vitamins, minerals, and phytonutrients, making this a powerful plant. A few of the benefits of moringa are reducing swelling, protecting the liver, treating upset stomach, improving eye health, balancing blood sugar levels, protecting the brain, and so much more.

Soursop is one delicious fruit. It is a tropical fruit, so it's not readily available everywhere and a lot of people have never even heard of it. One of Bobby's favorite teas is soursop tea. Soursop is known to kill bacteria, reduce inflammation, help with anxiety, and prevent and fight cancer. We do not eat the leaves or use them in any way other than making tea.

Blue butterfly pea is one of the most beautiful flowers I've ever seen. It's a vining plant and it loves hot and humid temperatures. We use blue butterfly pea in our soap products and to make tea or lemonade. It's my favorite summertime drink. Our bees also seem to love the flowers.

I call blue butterfly pea magical. When I boil the flowers, the water will turn blue. Then I'll add my lemon juice to the tea, and it will turn purple, just like magic. I've seen people make rice with the blue butterfly pea flower, turning the rice into this beautiful blue color.

Blue butterfly pea is known to promote healthy weight loss, stabilize blood sugar levels, support skin and hair health, reduce inflammation, and aid in heart health.

ADVENTURES IN HOMESTEADING

Have you ever smelled fresh lemongrass? The scent alone makes my heart dance. Lemongrass is my favorite herb to make essential oils with. I think our entire house is happy when we are extracting oils

lemongrass oil. Lemongrass is used to cook with and to make teas, tinctures, candles, fragrances, repel mosquitoes, cosmetics, and more.

Lemongrass is a mood booster, due to its wonderful scent, but it's also good to help prevent the growth of some bacteria and yeast, relieve pain, reduce swelling, reduce fever, improve blood sugar levels and cholesterol, stimulate the uterus, offer antioxidant properties, treat anemia, remove toxins, and burn fat, just to name a few benefits.

Mints are another happy mood herb. We grow a variety of mints including lemon balm, peppermint, spearmint, banana mint, pineapple mint, orange mint, mojito mint, and chocolate mint. They are some of the easiest plants to grow but very invasive. Growing mint in a pot is best to try to keep it contained.

Our chickens, quail, and rabbits love to eat mint. It's a very healthy treat for them. I sometimes extract the

ADVENTURES IN HOMESTEADING

from our mint plants for essential oils. We also make tea and seasonings with our mints. Sometimes when I'm out messing around in the yard, I pick a piece and chew it for a little while. What a refreshing herb!

Fresh mint makes some of the best herbal mouthwash. It's so simple: mint, non-chlorinated water, salt, and baking soda. I'll let it sit a day then pull the mint out and it's ready for use. We also grow mint all around our property to help deter bugs and pests. The squirrels, snakes, and birds don't like the smell, which is another thing that makes me happy about growing mint.

We all know how important brain health is, especially as we age. Gotu Kola is an excellent herb for brain health and it's so easy to grow. You can make an amazing paste for wound healing, tonics, tincture, anti-inflammatory, and so much more. I make tea every week so we can sip often during the week to help us focus and keep our brains healthy. The plant is so cute, the leaves look like little lily pads.

Gotu kola is native to India and the southern US and thrives in tropical/subtropical weather. If you are ever feeling a little anxious, take a sip of this amazing tea and it will soothe the anxiety real fast.

Let's talk about those uncomfortable hot flashes! My beautiful Soil Sistas know what they are! Black cohosh is the herb for that. We usually have menstrual issues when our estrogen levels are low. Black cohosh helps with inflammation and rheumatoid arthritis and also has sedative properties. I always tell women to make sure you are getting some probiotics in when taking black cohosh, but it's a great herb to check into and have around for those uncomfortable moments.

I love talking about turmeric and the different varieties we grow. My favorite is the black turmeric, which holds the most benefits of the turmeric family because it has the most curcumin. It's disgusting but I respect the benefits. We also grow white turmeric and orange turmeric. Turmeric is great for the skin, has anti-inflammatory properties, lowers cholesterol, is antimicrobial, and has anti-aging properties.

Turmeric is very popular in Indian foods and is found in a lot of medicines. It's a key remedy for many chronic health issues such as allergies, arthritis, diabetes, and psoriasis.

Lavender is another one of those happy perennials you want to have around. This beautiful purple flower, which puts off the most beautiful smell, just makes everything better. The aroma alone is a natural antidepressant that tends to relieve anxiety and stress. The oils can be extracted from the flower to use in cooking, massage, tinctures, and infusions.

Our nervous system reacts very well with lavender, leaving us happy and calm. It's also good for digestion, asthma, aches and pains, bites and stings, earaches, insomnia, and headaches. What a powerful plant, and again, all natural. I love it!

Tea tree is a mold-smelling herb that also heals the soul. I love using tea tree oil in my cleaning products, cosmetics products, and infusions. It carries antibacterial, antiseptic, antifungal, and antiviral properties, and it's an immune stimulant. It's a powerhouse!

We usually only use tea tree oil topically. It can be mixed in with your body wash and oils, or you can rub a little on any areas that may be having issues. It's great for chronic infections and can be taken with water to help you heal, but remember, a little bit goes a long way. So, you can mix about three to four drops in a glass of water to consume.

ADVENTURES IN HOMESTEADING

Oh, holy basil! Basil can take over your growing space, but it smells so beautiful. I know I say that a lot, but the smell of herbs just makes me so happy. If you are ever feeling like you have any respiratory issues, feverish, stressed, suffering from high blood pressure, diabetes, insect bites, ringworm, and so much more, you may want to check out holy basil. Our bees love it as much as we do.

You can make juice or tea for skin infections and rash, decoctions for fevers, and powders to help heal ulcers and sore areas. In the culinary field, they refer to holy basil as sweet basil.

When you suffer from high blood pressure, sometimes the doctor will prescribe you a diuretic. We drink cornsilk tea. Corn silk can reduce kidney stones, cure bladder infections, and treat bruising, swelling, sores, and boils. I trashed corn silks for years until I stumbled upon the information of how beneficial it is for our bodies, and our chickens and rabbits enjoy it too.

Let's take a trip back down memory lane. Remember when you would have a tummy ache or caught. a cold? The first thing I remember my mom doing was giving us room-temperature ginger ale. It took me forty-five years to learn that ginger ale probably didn't help me feel better or heal. However, once I learned how to make ginger ale with three simple ingredients, my life forever changed.

Jamaican hibiscus, also known as sorrel or rosella, I also call our blood pressure medication. I remember the first year we grew this plant, I did a little research and thought I read this plant would get four to five feet tall and bushy. Ha! I dropped a seed in my tea bed near the front door of our home and it grew about fifteen feet tall and six or seven feet wide. It was huge

but gorgeous.

This plant will start flowering usually around the beginning of fall and one tree will give up right until the first frost. I've seen people making jams and jelly with the hibiscus, but we just make tea. One cup of hibiscus tea daily, sweetened with honey or agave, can keep you off synthetic medicine for life. I always say diet and exercise are

ADVENTURES IN HOMESTEADING

important also. It's a delicious tea, though, and very beneficial.

A few benefits of using sorrel or Jamaican hibiscus are liver support, lowering blood pressure and cholesterol, promoting healthy weight loss, fighting bacteria, and providing antioxidants and vitamin C. This tree loves warm temperatures and adds a ton of beauty to your property.

Now I'm not one to be messing around in the woods or grassy areas, but I like foraging. One day while riding around Ossabaw Island, one of our friends was talking to Bobby about how to spot mushrooms. Chicken of the Woods really does taste like chicken! We found a few good pieces and brought them home. Bobby cleaned them up, and I promise you, it was just like eating a piece of chicken. So, I started learning about other plants that grew wild around our area.

Goldenrods are beautiful bright yellow flowers. A lot of people get it mixed up with ragweed. Goldenrod carries so many benefits, and it's free.

I love making homemade healing products. Every year, I make fire cider for my family. I start by cutting up an orange and a lemon into wedges with the peel, chop an onion, fresh garlic, ginger, horseradish, cinnamon, turmeric, cayenne pepper, dried elderberry, and whole black pepper, and drop everything in a half-gallon jar. I'll then add apple cider vinegar that I make with apple scraps, filling the jar to the top. I place the top on it and give it a good shake. I'll then set the jar in a dark cool space for six to eight weeks, occasionally shaking my jar when I think about it.

After a few weeks, I'll strain my solids out and give them to the chickens as a snack and add about a cup of honey to my liquids. I mix it well, then we have a natural preventative to take daily until winter to help keep us well. I call it our natural antibiotic; you can feel the wellness going down with every swallow.

I turned our China cabinet into our apothecary. I love that everything can be stored in that one piece of furniture. I have our freeze-dried jarred herbs on the top so I can see everything from the window. There's a drawer where I can store all my labels, notebooks, and other things. Then there's space at the bottom of the cabinet where I can store extra jars, herbs, tinctures, and other supplies. It's just so convenient and organized, and I love organization.

Holistic healing can be a little intimidating, but herbs can't hurt you and consistency is key. I do a deep dive into every herb I run across, one at a time because there are so many amazing herbs and plants, and it can be a bit overwhelming in the beginning. You can check out our videos about some of the herbs we grow at www.youtube.com/watch?v=hzUvLrZOKUc

Homesteading Housewife

Sherrie

As a homesteading housewife, organization and sticking to a schedule are important for me. My days are pretty routine, but I leave space in them just in case. My job is to cook, clean, and look after the animals. Bobby and I take care of the plants together. We are also content creators, so I keep our social media sites active and respond for the most part, taking orders, shipping, and filming a lot of my day.

I'm usually up at 4:00-4:30 a.m., packing Bobby's lunch and seeing him off to work. I'll lie down for an hour or so, then I'm up for the day. I head to the bathroom to get myself together and then feed the dogs. I tidy up the kitchen and prepare for dinner by getting the meat out of the freezer and setting everything up so I can cook around noon. If I'm baking dessert, I'll go ahead and get it done, usually by 8:00 or 9:00 am. I'm not a breakfast person, so I'll just go on about my day, filming and playing in the kitchen. If Bobby is off, I'll fix his breakfast. Then I'll check on the chickens, rabbits, and quail and walk the yard to see how the plants are doing.

BOBBY AND SHERRIE BLACK

I set aside one hour three times during the day to check out social media, Facebook, YouTube, and Instagram. We like to respond to as many comments as we can and interact with our followers.

When we began creating content for YouTube, a lot of the people whose videos we watched talked about growing with seeds. We didn't understand why it was such a big deal. Being impatient and wanting to hurry up and get fruit and vegetables, we'd rather buy starts to save some time. Then someone sent us a package of seeds. Most of these seeds we'd never heard of. So we began to research. I never knew there were so many varieties of tomatoes. We figured there were maybe two or three varieties, like the basic slicer and Roma tomato, that's it. But no: there are so many varieties, with beautiful colors and different tastes, so we jumped in and learned all we could about seed starting.

Favorite Plants

Sherrie

I remember Gt Jr grows it Alaska sending us jicama seeds when we first met him on YouTube. I'd never heard of jicama, but I figured we'd drop a few seeds and see what happened. I read that jicama seeds were poisonous, so I wanted to keep them away from the chickens and dogs. I found an empty pot of soil in the back of our property, dropped a few seeds in it, and wet the soil. Then I walked away and pretty much forgot about it for a few months.

One day Bobby was out playing with his bees, and I thought about the jicama. The vines had grown, and the seed pods were dry and cracking open. So I collected the seeds and pulled up the vines. To our surprise, there were big tubers under the soil! I didn't do anything to this pot and this was the result! Amazing! I pulled all the vines and found we had six beautiful round jicama.

Then it was time to figure out what to do with them. We found that some people would just slice them thin and eat them raw or on a salad, while others prefer them sliced like French Fries and fried. So I did both, fried for Bobby and on a salad for myself, and they were delicious. A seed created all this deliciousness. This taught me a valuable lesson about seeds. They are small but mighty.

Another favorite for me is the Taiwan long bean and snake bean. I love things that grow big for some reason. However, they are also delicious vegetables.

BOBBY AND SHERRIE BLACK

The snake bean grows so long and thick that it looks just like a huge snake. Bobby also loves growing beans and peas. He said it's the shelling part he likes the most.

Watermelon and melons are also one of our favorites. Who would have known there were so many varieties of them?

I remember my late uncle introducing us to yellow meat watermelon. For most of my life, I only knew of one red watermelon. Now they have a yellow rind, with red insides, orange, pink...they all taste different, but so good.

The chickens love watermelon, so we share some with them, and we make watermelon rind candy and pickle some of the rinds too. We have a zero-waste policy. What we can't eat, the chickens and rabbits can have, and we will compost the rest. It's so funny to walk around the chicken coop and see seedlings popping up of watermelon, tomatoes, and other plants we had given them over the years.

Hurricane Season

Sherrie

Hurricane season is always a tricky time for us. We have been blessed to not have any direct hits or major damage. We did lose our greenhouse once, but that's another story. We grow a lot of our tropical plants in pots to protect them from the winter. As soon as we hear a storm is coming, we spend the day pulling plants inside the garage and house, strapping down the rabbitry, and locking up anything that can fly away.

Bobby will stake the banana plants, especially if they have fruit hanging. The plants get so heavy, it doesn't take much wind to topple the plants over. It can be a lot of work but always worth it. We hunker down and ride the storms out. But thank God it's always nothing. All that work and most of the time the wind barely blows. But we'd rather be safe than sorry.

Social Media, Really?

Bobby

We wanted to show our family what we were doing. Neither one of us knew anyone in our personal lives who was doing anything like this, and we were eager to show our family. Since we were staying away from people due to the pandemic, the only way was social media. We posted the process on our personal pages. All this was so exciting.

We knew there were many people out there like us, who didn't grow up having this experience and needed to know about this wonderful lifestyle. So we decided to start a YouTube channel. Neither one of us knew anything about YouTube, so, as with any other thing we didn't know about, we did some research and took our shot. We were going to tell the world about this.

Except it didn't happen like that. We made a couple of videos, but they were horrible. So, we kept putting off going live.

We finally did it on Sherrie's birthday. This was hilarious looking back on it. At the time, it was terrifying.

ADVENTURES IN HOMESTEADING

Sherrie, the outgoing one of the two of us, was petrified, so I was a mess. I think we were just sitting there, staring at the camera. The whole thing suddenly felt invasive, like strangers were going to be coming into our lives and watching us. We decided not to wait too long before trying it again. We knew that if we waited, we'd never actually do it. Thanks to someone we met on YouTube, we got through it and got comfortable doing livestreams. (When I say "we," I mean Sherrie. It took me quite a bit longer.)

Community, Yeah

Bobby

To our surprise, there were so many farmers, homesteaders, and gardeners on YouTube there were communities already. How did we not know this? It was information overload. It was love overload. Sherrie and I jumped in with both feet running. We filmed our videos the best we could and unloaded a lot. We made guest appearances on every channel that would have us. We popped up in people's lives. We told everyone how we felt about this homesteading thing.

Not only did we find a community of likeminded people, but what we learned from that community helped us grow ourselves. Before getting on YouTube, we only grew plants from the starts we could find. Afterward, we found out about all the great varieties that we were only going to be able to grow from seeds, which has opened so many new possibilities.

Raised the Beds

Bobby

We sat down and decided that I had to build more beds because we had to grow all these amazing new vegetables. If you are going to homestead, you need to have some skills in creating, making, and developing things. If you don't have carpentry skills like me, you should start working on them.

Take every opportunity to build things that you need, even if you can afford to purchase them. Sherrie and I came up with ways to do this. We used a privacy fence to build raised beds. The privacy fence was eight feet long. Our beds were six feet long and four feet wide. This was six privacy fence panels and a one-inch by one-inch pole.

This cost us about twenty-one dollars. We built about five of them in total, at a cost of a little over one hundred dollars total. You can't buy one bed this size for twice that price.

Rotten Rottweiler

Sherrie

Nala, Thor, and Trudy Black are our dogs, our spoiled rotten babies. I remember picking Nala up when she was eight weeks old and a big ole chunky baby girl. She was so sweet and loved to play. Lovie would call her Lucky (her pet pig from a year ago). Potty training for Nala was a breeze. She slept in the bedroom with us until we found puppy number two. She's now three years old and a total Daddy's girl.

Thor came along when Nala was about six months old. Nala was so clingy she desperately needed a companion. We drove to Augusta to pick him up and boy was he cute. He was so tiny! Or Nala was just huge, ha! He had a huge head and a tiny body; I kept asking Bobby if something was wrong with him. Nope, he was perfect, eight weeks old and so calm and curious. We had to be careful with Nala and Thor because Nala was a big girl, and we didn't want her to hurt him. Potty training was a job in a half with him! It took weeks to get him to understand the bathroom was outside. But once he understood, he was good. He's a Momma's boy for life. I can't use the bathroom without my security guard.

We allowed Nala and Thor to breed at two and a half years of age, on December 1, 2022. We were so careful with Nala, checking her daily and watching for any signs of change. I wanted to document her pregnancy from day one. Call me the MidWoof because I didn't miss a thing. She was the perfect pregnant dog. We'd

ADVENTURES IN HOMESTEADING

take her to the vet to try counting the puppies, but there were so many we couldn't get an accurate count. She was healthy, huge, and very happy.

Finally, the day came, February 5, 2023. Nala spent the night in her whelping box Bobby built in our studio, and I was lying right beside her. She seemed very restless, digging in the box and walking around a lot. We stayed up all night and all the next day. Finally, around 4:00 p.m., I was so tired I laid down to take a nap, and I heard Nala making a weird noise. I quickly sat up and said, "Nala are you pushing?" She got up and went to get in her whelping box. So of course, I followed her to see what she was doing. Sure enough, she was pushing. I told Bobby she was ready so we both sat in the room with her for a few minutes. I could feel a puppy in the birth canal when I checked. We filmed the entire birth live via YouTube. You can check out Nala's puppies' birth at www.youtube.com/live/IF8lE--ygw8?si=bU1SOkKiQeTz1-od

Of the eleven puppies born, nine survived. The journey with Nala, Thor, and their nine babies was amazing. They are smart little cuties. I had these babies' potty trained at three weeks old! Consistency is key. "Potty over here!" They loved each other when I said that, and they would potty over there. Eight or nine weeks later, we rehomed all but one, who was so attached to us we couldn't let her go: our Trudy Girl, also known as Booty.

Let's Do the Bunny Hop

Bobby

While we were doing our YouTube thing, we learned about rabbits. This was something that we never considered or even knew anything about. We did our research like we do with anything and found that rabbit at one time was the dominant meat in America, beating out chicken.

So, we asked why this would be the case. We found that the creation of a famous cartoon character made people feel a certain way about eating rabbits. The vegan movement also contributed to slow down of the use of rabbit meat. While we respect vegans and vegetarians, we are meat eaters. We couldn't find a good reason not to raise rabbits. So, we consumed lots and lots of information about raising, harvesting, and cooking rabbits.

We found out that we could produce about six hundred pounds of meat a year with three rabbits, more than we could get with a cow, and it would cost a fraction of the money. Sherrie and I were extremely interested, so we searched our city for people who were raising rabbits.

Finally, we found an amazing local homesteader who had rabbits for sale. We were invited to come pick up some of our own.

We met Sandra and Hurbert, the first homesteaders we met in person. They had always lived this kind of life, which we thought was amazing. We talked a great deal about homesteading and the breathtaking

ADVENTURES IN HOMESTEADING

amount of work that they put into homesteading. We instantly fell in love with them and how genuine they were. We purchased one buck and two does and were excited to go home with them. But despite all our research, we hadn't put a lot of thought into housing the rabbits. I believed that I could just build a hutch quickly, but after seeing Sandra and Herbert's setup, realized that it may take longer than I thought. It just so happened they were willing to sell one of their homemade rabbit hutches. We rented a trailer because this thing was massive, about fourteen feet long.

The hutch was very sturdy, standing three feet off the ground, and it came with a watering and feeding system. We got it set up and it was just enough for the three rabbits we had.

Now this was an opportunity to build one on our own. So, using the one we purchased as a guide and a little ingenuity, we decided on a method. We went with lightweight material, eight-foot long one-inch by one-inch treated lumber.

We purchased about twelve one-by-ones and a roll of hardware cloth. I then started making the hutch only using the one-by-one treated wood, eight feet long and two feet tall. I divided the hutch into two compartments, four feet each. We wrapped the hutch with hardware cloth. We made doors and secured the hutch. We then built a stand from some two-by-fours that we had. The stand was three feet tall and built to hold the hutch. In total, we spent about one hundred dollars on materials.

Soon after, we bred our bunnies. We knew that we needed to take the doe to the buck and never the other way around. The first time, we were startled when it looked as if our buck suddenly passed away.

BOBBY AND SHERRIE BLACK

Each time the buck mounted the doe, he would seize up and fall on his side like he had died. This is called the fall-off. After waking up he was back trying again. After initially being a little frightening, it became quite funny.

About thirty days later, we were having kits (baby bunnies). Rabbits can be bred right after giving birth. In a situation where they are not separated, this is what would happen naturally. We decided not to breed so often. After our doe weaned her kits, we gave her a little time to herself.

One day, I was at work, and my wife was in a panic because she was alone while one of the does was giving birth, and it was Sherrie's first time. Luckily, she was able to get ahold of a friend who was familiar with the process. She calmed down and everything was fine.

A couple of the kits did not make it, but we were told that was normal. After about three weeks, we separated the kits into a grow-out pen so the mothers could get a break and get ready to be bred again. In about eight to ten weeks after being born, they were ready to be processed. I am the one that does the processing, so we decided that I would not interact much with the rabbits.

Sherrie did not join me in this process, and we did it like that so it could be easier for both of us. Don't feel bad if this is hard to do, because it should be. We give all our animals the most amazing life, feeding them well, making sure they are safe and Sherrie loving on them. I decided on the brome stick method.

I knew doing this wrong would cause pain, so I practiced dispatching them many times because I did not want them to suffer. I successfully was able to put fifteen rabbits in the freezer. Now I had to figure

ADVENTURES IN HOMESTEADING

out how to cook them so my wife and I wouldn't feel like this was a mistake.

I decided that I would smother the rabbit in the same fashion that I do chicken or pork chops. Sherrie loves it when I smother chicken and pork. I figured this could not fail.

So, I seasoned the rabbit after quartering it up. I had the two front quarters, the two rear quarters, and the tenderloin. I then seasoned my flour and shook the meat in the flour, making sure that they were well-coated. Then I slowly fried them in a small amount of coconut oil. After they were close to being done, I removed the meat.

I put my diced onions and peppers in the pan with some butter, making sure to keep all the fried leftover (crumbs) of rabbit in the pan. I sautéed my onions and peppers with the bits from the rabbit. After my onions were translucent, I started to slowly incorporate my liquids in the pan. I decided to use some turkey broth that we made from our last Thanksgiving turkey.

You can use any broth you like or water. As my liquids incorporated all the flavors from everything in the pan, I placed my meat back into the pan. I then added more flour and a little cornstarch. As the temperature rose, I poured the rest inside the pot as a thickening agent. Then I covered it and let it slow cook until it is at the desired tenderness. Flavor on top of flavor, how could this miss?

White rice gravy rabbit and peas was our test meal. Score! We knew from this point forward that rabbits were going to be a part of our homestead. All I heard from Sherrie was, "Poor Bugs Bunny," as she was sucking on the last of her meat, laughing. We have eaten rabbit many times since in different ways, and they all are delicious. Cooking in liquids (stewing,

BOBBY AND SHERRIE BLACK

smothering) seems to be the most enjoyable for us.

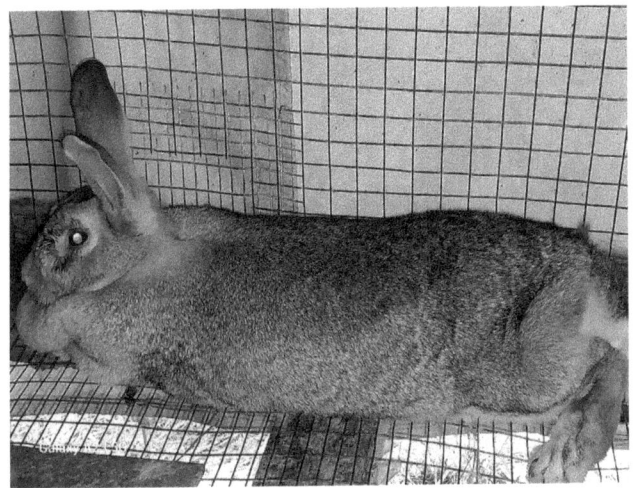

Sherrie

I remember Bobby mentioning raising rabbits! He went down that rabbit hole so fast and deep. We ran across someone not far from our homestead who sold rabbits. As soon as I saw the girls, one red and one black-and-white, I already had names. Then when we saw the male, who was thumping his hind legs, and I knew this was

Thumper!

We named our girls after our sweet friends Sandy and LaLa! Our beloved friend Sandy passed away on her 108th birthday while living with Bobby and me. The rabbit had snow-white hair and she reminded me of Sandy so much, sweet and quiet but observing everything and everyone.

Our red bunny made me instantly think of my "bestie boo," LaLa, short for Louisa, strong, smart, and vibrant! I think between Sandy and LaLa, a lot of

ADVENTURES IN HOMESTEADING

my inner homesteading being came to life. I used to make chemical-free cleaning products and personal hygiene products for LaLa years ago.

Just for fun, her birthday and Christmas presents always came from our homemade products. Along with our boy Thumper, these were our original breeding pair of New Zealand rabbits. They are perfect.

Sometimes we'd forget to latch the hutch and the rabbits would get out. They'd be out all night, and I would freak out every time. Luckily, they'd be sitting right under the hutch waiting to be put back. So we would remind ourselves to double-check the locks. The rabbits did not want to be on the ground; they loved their homes, and they were safe there.

Rabbit is our main source of meat. Between the pair, we could easily produce up to six hundred pounds of meat per rabbit. That's more than enough for Bobby and me. However, we don't breed our girls that heavily. They love the winter and don't care for the summer much. In the summer we let them have a break from breeding.

During the fall and winter, we will breed the girls every six to eight weeks or so. Rabbits can breed every thirty days, which is a lot. If push comes to shove, we will breed them heavier, but for now, breeding in the fall and winter is more than enough.

Rabbits also make the best fertilizer. It's considered cold compost, and we can use the manure directly in the garden. Bobby saves the pelt or fur to tan, so in case we ever need warmth, we are covered.

BOBBY AND SHERRIE BLACK

Kitchen Scientist

Sherrie

I love playing in the kitchen! I've done a lot of research on our food system and I'm not too happy with what I've seen. What is organic? I know what we consider organic to be, but what does the system consider organic? What are the chemicals and preservatives doing to our bodies? I have so many questions!

So, I started looking at simplifying our recipes. Not only is homegrown better, but homemade is also the absolute best. We don't use preservatives, so the food doesn't last long, but it's okay; it's safer for us this way.

I don't want my family consuming stressed-out meat either. If we consume meat that comes from animals that had a stressful life, we are adding that stress to our bodies.

I've learned a lot about the foods we consume, so when we grow them, we know all we can. Growing and raising our own is one of the best decisions we've ever made. Overall, we feel a lot better and the peace that comes with this lifestyle is priceless.

Homesteading

Sherrie

This homesteading journey has been amazing! We've learned so much and have so much more left to learn. Over the last three years, we have grown closer as we gain these much-needed simple life skills. We have learned so much from each other. Yet there is always more to learn, and I always ask, "What's next, God?"

I'd love to upgrade our space and have a full yard. My dream animal is a miniature cow. They are so cute! The joy that our animals bring to our homestead makes me want more. Homesteading on our one-acre suburban homestead is perfect. We can keep up with everything a lot easier than having multiple acres, though it's still a lot of work.

Can we handle more? Of course we can. However, this works perfectly for us and we can help provide fresh food for others, which makes us feel so good. When I think about having multiple acres, the fear of it being too much is there. We love being able to use our entire acre to grow and raise our food. But in my heart, I know God will guide and direct us exactly where we need to be and when.

For more information on Black's Tropical Homestead and our amazing lifestyle, check out our YouTube channel at www.youtube.com/c/blackstropicalhomestead

SoilFamily Expo

Bobby

We got so close with the people in our community that we decided that we were going to have a little meetup on the river. Through Sherrie, we were able to connect with our local children's museum in Savannah.

One of their missions is growing food and educating our youth on healthy eating. We spoke with them, and they were willing to host our meet-up. I was a little stressed out, as I'd thought we were going to have a meet-up on the river, just a few of us, and have dinner. Now we had a venue and vendors, and we went from meeting on a single Saturday to an entire weekend.

It was up to me and Sherrie to make sure this went off, and I did not feel ready for the stress of being responsible for all these people coming to town, wanting them to feel comfortable, and wanting them to feel like it was worth it. The first one was amazing. We had content creators and supporters come long distances to meet us for the very first time, people who had known each other online for years. We were amazed that these were all good people to the core. How could they all be good? How could they all be what they seemed to be? Our hearts were full.

We had people join us from California, New York, Maine, Texas, Indiana, South Carolina, North Carolina, Georgia, Florida and more. We laughed, played, hugged, and ate, and when it was time to go, we cried. It was amazing. Lead Farmer showed up.

BOBBY AND SHERRIE BLACK

Lead had been an influencer for eleven years up to this point. Most of the people there were very familiar with Lead Farmer, and he was the reason some of them became influencers. So his being there was an added treat.

Thanks to one of our Soil Sistas, we were able to get a proclamation from the City of Savannah and Chatham County for the SoilFamily Expo. How did this happen? We'd told a few people, but more people needed to know that homesteading can make your life, your family life, and the world so much better. Sherrie and I spent every vacation and all our spare time visiting nurseries, gardens, and farms. We could not gain enough knowledge or experience. We were going to tell everyone.

Sherrie and I decided we were going to travel to Virginia to go to Homesteaders of America. We noticed when we arrived that were thousands of people there, including successful content creators. All I could think of was how amazing it must be to do this thing that I now love for a living. I know that these people are looked at as celebrities, and the last thing I wanted to do is be that weird person that they must get tired of dealing with.

But we noticed that they were expecting to be approached, so we had a light conversation with a lot of them and continued on our way. Surprisingly, people were approaching us and knew us from our own content. We were shocked. We thought we were nobody in this world of content creating, but it was amazing.

Also, even though we were two of only about ten people of color we saw present, there were no uncomfortable moments. We're not sure if the people living this way is what makes the lifestyle so great or if the lifestyle

ADVENTURES IN HOMESTEADING

makes people great, but either way, these are our people. Race, politics, religion, sex, age, height, weight, whatever: none of it matters. The only thing that matters is being and doing the best they can with what they have. These people are SoilFamily.

SundayBackyardFarmer, Just Dorsha, Kinfolks Farm(Shane and Kim), Black's Tropical Homestead (Sherrie and Bobby Black)

Top row left to right Broke Farmer, Sunday BackYard Farmer, Cassandra South Fulton Gardener, Best Yet Journey, Farmer Q, Spoon-N-With Sunshine, Food By Faith Garden To Table, Thyme To Grow, Sherrie Black, Just Dorsha, V and Family, Cali Homesteading With PoohBear and Anne Dale Homestead

BOBBY AND SHERRIE BLACK

ADVENTURES IN HOMESTEADING

To Bee or Not to Bee

Bobby

We met so many incredible people that weekend that it is hard to believe, including an amazing family that we met on YouTube, Jarred and Angela of The B-Hive Homestead. We spoke with them online. We then met up with them at Homesteaders of America, so we knew they were good people.

Jarred knew I wanted to get into bees and said that he would have a hive waiting for me. After the convention, we went to The B-Hive Homestead. I was extremely excited. When we got there, we were welcomed so amazingly by Jarred and Angela and their two beautiful kids. After the chicken and cows, we finally got around to the bees.

The enthusiasm and knowledge that Jarred had for bees was incredible. It drove my hunger for knowledge even more. Even the babies knew and were not afraid of the twenty-three hives he had on site.

Jarred was happy to answer all my questions, and they brought Sherrie and me a lot of different types of honey to try. They were a variety of different shades, from dark brown to yellow, and they all had strong differences in tastes, from molasses to a light floral taste and everything in between. I'd never known honey could have such a variety of flavors. After all, the honey sold in grocery stores all look the same. Jarred explained that these honeys are harvested at different times of year, and some were harvested from different areas, which explained the different colors

ADVENTURES IN HOMESTEADING

and tastes.

Jarred gave me three boxes of bees and everything I needed to get started with beekeeping. I bought a suit from Homesteaders of America, and we transported all these bees from Virginia to Georgia while stopping in North Carolina to see Anne Dale Homestead (Land of Plenty) on the way home.

Amusingly, the bees were able to get out of the hive and were flying around the car. I was comfortable because Jarred and six-year son Maliki explained that the bees won't sting you if you don't swat or slap at them. On the other hand, Sherrie was not convinced and freaked out a little bit. We had a ball with Anne Dale and Worth (Anne Dale's husband) we did a livestream at the Land of Plenty on YouTube. Anne Dale and Worth fed us a great meal and we got on our way. It was an excellent weekend.

This began our journey in beekeeping. I consumed as much information as I could, but I soon discovered that all beekeeping situations are unique. There are basics but you can't follow anyone's exact experience. At first, Sherrie didn't plan to work with the bees, so I did most of the research, but I uncovered so many questions that I felt I needed to have her working with me. We got her a bee suit and it was on. We joined the local bee club, where we could hear the experiences of local beekeepers, which is more useful than those of beekeepers in other cities and states. Getting to know local beekeepers was also wise in case we ever needed to help each other in an emergency bee-raising situation.

We were able to harvest more than twenty gallons of honey in our first season on our homestead. We owe that success to the B-Hive Homestead, our bee club (Coastal Empire Beekeeping Association), and the rest of our community.

This Little Piggy

Sherrie

I always wanted a pig. Bobby thought I was completely crazy. Then one day he said the magic words: "I think we should raise pigs for meat." We did a deep dive into learning how to raise pigs and what the processing is like.

Could we do this here on our one-acre suburban homestead? We drove to Kinfolk Farms, which was a dream come true. I was so excited—we were about to become pig parents! —but had to keep calm. We arrived at Shane and Kim's farm and spent the day with them, helping with the pigs and enjoying lunch together with a few other friends.

The ride home was funny, as we had pigs in the back of our car. This was becoming routine, bringing animals home in our backseat. The smell, however, was another story. Applewood and Bacon were pretty small, so they could fit inside a dog crate, and they pooped that thing up!

Finally, we arrived home, and they went straight to the pig pen my husband built. One of the pigs was so small it could slip right through the cattle panels, and suddenly it was out. We started to panic because one of our pigs was out and could be running all over the neighborhood, a suburban neighborhood at that! So, Bobby quickly ran around the neighbor's house to chase her back toward the pen, and she slipped right back inside. So, then we had to figure something else out so they couldn't get out that night.

BOBBY AND SHERRIE BLACK

I suggested putting pallets up: they shouldn't be able to get past them, but it was dark, and we had to move fast. Bobby drove up the street to Dollar General and found a few. We built another fence with the pallets and finally we could go inside, shower, eat, and rest. The pigs settled in quickly. I spent a lot of time with them, sitting in the pen every day for hours. I'd go live on YouTube to share our adventures with pigs in our one-acre suburban homestead. Is this doable? It is... however, not in our suburban neighborhood.

Bobby

Should you try and raise pigs in our situation? Let's talk about it. One of the amazing farms that Sherrie and I visited was Kinfolk Farm. Shane and Kim have a beautiful farm where they raise pigs. After seeing them, we wondered if we could raise pigs on our homestead without disrupting our neighbors.

I thought, due to the way our yard was set up, that we may be able to raise them, and no one would even

ADVENTURES IN HOMESTEADING

know they were there. While at Shane's and Kim's Farm, I didn't notice any bad smell, and the pigs looked as though they were tame. This was our first time being around farmed pigs.

A few months later, Shane mentioned that his pigs had given birth, and he would be looking to get rid of some. We did our research and decided to get two. I built an enclosed area in the rear of the property that was thirty feet by twenty feet using a hog panel. I put electrical fencing all around this area. I was able to get a fifty-five-gallon drum for water to which I attached a nipple.

We got a large pan that we were going to use for food. We planned our trip and off we went to get our pigs. When arrived Dorsha (Just Dorsha) and Nick (Sunday Back Yard Farmer) met us there. We walked the property, learning about the pigs and the muscadines grapes that he grows and walking the beautiful lake on the property. Shane and Kim fed us a great meal.

Then it was time for us to get on our way back to the homestead. We were not sure how our newly acquired pigs were going to travel. We had them contained in a dog kennel. The thought of them getting out and running around in the back of our SUV was not pleasant. We made it back to the homestead without incident and put them in their pen.

We made the mistake of not turning on the electrical fence before putting them in or immediately after. Something else I did not consider was there were some spots in the panel that the piglets could squeeze through. One of them did just that, and the chase was on. I worried I might not be able to catch her, and it would be my fault that a pig was roaming around our neighborhood. But after a chase, I caught her. Thanks to some hands-on experience at Shane and

BOBBY AND SHERRIE BLACK

Kim's farm, I knew how to handle them.

We were now raising pigs on our homestead. Over the weeks they seemed to double if not triple in size. Sherrie was amazed and spent a lot of time out there with them. Because we want to have all the skills needed to be self-sufficient, we decided to do processing ourselves. We purchased books and searched the internet for instructions. Then the perfect storm happened. We never considered the amount of rooting that they would do.

They were able to turn this elevated area in which I had them into a depression. The bigger they got, the worse they dug. Then out of nowhere, we got the most rain I think we've ever gotten. All this rain pooled in the pen. The pigs loved it, but it turned a natural smell into a very bad smell. Poop and water mixed together, as it turns out, don't smell too good. Then we had wind that blew that smell all around. If we were on land with no neighbors, we would have been able to deal with this, but because we were not, we were uncomfortable, not wanting to bother anyone.

We had to process our pigs immediately. At this point, they were only about two hundred pounds each. We were hoping for them to be about three hundred at processing time. We were not prepared, but it had to be done soon. Using some four-by-four-by-eights and my trees, I made a host. I was very careful about how I dispatched them, knowing if I did it wrong, it was going to cause them pain, and that was the last thing we wanted. I got some strong rope that I used to hoist them up. Then we were able to clean and skin them. Sherrie was a mess before it started, and I didn't expect her to be a part of this process, but she pulled it together. I continued to watch Sherrie without her knowing to make sure she was okay. I never saw any sign that she was in trouble or couldn't do it. She was

ADVENTURES IN HOMESTEADING

incredible, and I told her so. We got it done. We took the meat into the house in halves.

I then put all I learned into play. I was able to get all the cuts from all the right places thanks to the butchering book I'd bought. As I butchered, Sherrie wrapped and labeled all the cuts using butcher's paper. We were no longer raising pigs, but we had meat that would last us for some time. If someone were to ask if they should raise them in a situation like ours, I would say no. If you are in a rural area, however, then I'd say go for it. I think being able to raise your food, no matter what it is, is the best way to go.

SoilFamily Wow

Sherrie

What started as a little meet-up around the fountain downtown ended up being the event of the year. Bobby and I quickly noticed the importance of community. About six months after we started creating content for YouTube, SoilFamily Expo was born.

We decided to host this yearly event for people all over the world to come together in real life. You don't have to be a content creator, gardener, homesteader, or anything, just good, like-minded people coming together to help build community and fight food insecurity. The Chatham County Commissioner presented us with a SoilFamily Day Proclamation. What an honor!

SoilFamily Expo Inc. is a 501(c)3 my husband and I created in 2022. It's a beautiful time in Savannah, Georgia, getting together with our amazing Soil Brothas and Soil Sistas, dancing, teaching, learning, sharing a family-style meal, and creating beautiful memories together. It amazes me how far people are willing to travel to spend the weekend with us, even as far as California. With a host of sponsors every year, SoilFamily Expo gets bigger and better.

ADVENTURES IN HOMESTEADING

Bobby

At the time of this writing, it's time for the second SoilFamily Expo. It looked as if it was going to grow from last year, and my stress level was going to grow too. Trying to get donations is an enormous task. Putting everything in place is a struggle. Who works this hard for no money?

Well, we do! We know if we can pull it off, it would mean so much to so many. That in itself is worth it all. No one knows what goes into all of this, and they don't need to know. All we would like is for everyone to know that they are loved and appreciated for being part of the family. We want those who feel like they don't belong in their small circle to know that they matter in this one. All we require is for you to be a good person. Incredibly, our max of one hundred people has registered for dinner. Now we know it's going to be way larger than last year's. The experience was amazing, my wife put together some things for the early arrivers to do to experience some of the city. We managed to get enough funding to get entertainment. We had so much fun meeting all our new family members. We danced together, laughed together, and had a ball.

The expo the next day was amazing, and even more people came. After the expo and dinner, it was back downtown, where we danced the night away. Our farewell breakfast was full of tears for it to be over. The amount of love that we have for so many people in all these pockets all over the world is incredible. Never believe you must be a hateful, mean person to be respected.

BOBBY AND SHERRIE BLACK

People will love the wonderful person you naturally are. People like you and me are the only ones that can save this world. We love you for doing the work.

Freeze Drying Fun

Sherrie

One day, I was walking around the house cleaning and passed our China cabinet. What a waste of space here, I thought to myself. We never use China; we don't even look at it. So I cleaned it out, purchased mason jars and labels, and started filling the jars with our herbs. In the beginning, I would hang my herbs to dry, which takes almost forever, then I found out about freezer dryers. I knew immediately I had to convince Bobby we needed a freeze dryer. It took a few months, but finally, he gave in.

Now we can preserve our herbs and produce them for up to twenty-five years! Freeze dryers are an investment, but they will pay off in no time. Zero waste, from our tropical fruit, herbs, ice cream, candies, dessert, meat, dairy...anything can be freeze-dried, except honey, which is no problem because honey does not expire!

Just think what if? What if we can't get out of the house or to the store for some unforeseen reason? What if we have entire meals and drinks stored safely away for our family and friends to eat? What a blessing!

Bobby, a.k.a. Dancing Banana Man, grows thousands of bananas every year, but how can we preserve them for the winter? We can freeze dry them, and they'll hold ninety percent of the food's nutritional value without losing color, flavor, or smell!

I learned to can from a few of our YouTube friends,

but I was still afraid of it. We also worried about the power going out and the jars heating up too much. Is the food safe to consume? I still haven't gotten over that fear of eating our canned goods, but with the freeze dryer, we are in great shape!

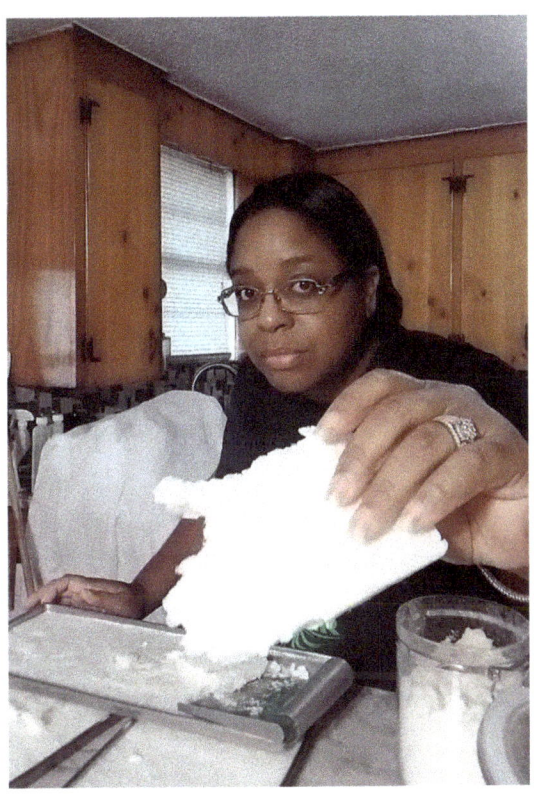

The Summer of Help

Bobby

It was a beautiful day. I was all hyped up ready to go work out. I looked over at Sherrie and she had this cat-that-ate-a-mouse look on her face. I looked at her and waited for her to tell me what was on her mind. Then she said, "Babe, I applied to go help Justin Rhodes help other homesteaders on their homestead. I didn't think they would pick us to help, but they did."

I said, "What help? Who? Where?"

Sherrie replied, "I don't know who, but it's in North Carolina. If you don't want to, I'll email them back and cancel." She then said it would be over the coming weekend, starting Friday morning.

I was hesitant. I needed to work. We would be going all that way and spending money instead of making money. What would this mean for our summer Florida trips? I told her I didn't think it's enough time to prepare, and she said, "Okay."

I went to the gym, which is where I do my best thinking, and I started to come around to the idea. Even though we would be going into this blind, I thought how good we would feel if we could give someone a hand. I thought we could also learn some things because there is still so much to learn. I could just cancel any work I had for Friday, and we could get up early on Friday and just go. I called Sherrie while I was at the gym and said let's go.

BOBBY AND SHERRIE BLACK

Sherrie, in her typical nonchalant excited-but-not-going-to-show-it fashion, said, "Okay."

Sherrie was emailed an address and time, but we knew nothing else. We just assumed that we would be joining a lot of other people who applied to come. We did not know if Justin would be there or if some employees would be directing everyone on what to do. Either way, we were there to give back to a lifestyle that we had learned to love. Sherrie and I got up around 3:00 a.m. to head to North Carolina.

\Our goal was to travel, and Sherrie was in awe of all of the beautiful North Carolina mountains. I'm from Savannah and it's very flat, so to see all this scenery was breathtaking. Our poor vehicle was in shock, I'm sure, traveling in the mountains. Before we knew it, the GPS was saying we had arrived.

We pulled up to a beautiful home on a corner lot. The first person I saw was Justin, removing tools from the back of a truck. I said to Sherrie, "There goes Justin." Then I added, "Is that The Fit Farmer?" Sherrie confirmed it was. (While we are content creators, Sherrie is the one who consumes most of the content. I am only familiar with those that everyone knows.)

I got out and went to help get the tools out of the truck. I spoke to Justin and Mike. The homeowner came out, introduced herself to us as Bri, and thanked us for coming to help. Justin said, "Y'all don't know who Bri is?" I looked at Sherrie, a little panicked, wondering if we were supposed to know. She looked back at me with a similarly uncertain expression.

Justin said, "Wow, y'all came to help and didn't even know who y'all was helping." I thought to myself, That's the idea, right? Now let's not be weird and ask who she is and get to work. And so, we did.

ADVENTURES IN HOMESTEADING

While working I noticed that Sherrie was talking to Anne Dale on the phone, as she does almost daily. We didn't tell anyone what we were doing because we didn't know if we should, but while we were there, Sherrie went ahead and told Anne Dale, who asked if it was Bri From Scratch. Sherrie confirmed that her name was Bri. Anne Dale emailed a picture and asked Sherrie if this was her, and Sherrie said it was. Anne Dale was excited because she was a follower of Bri.

Sherrie walked up to me and said the homeowner had a YouTube channel, and not a small one: she had some two hundred thousand subscribers. That definitely explained why we should have known who she was.

We also met Brea and Ben (Growing Up Holt). To my surprise, Mike and his family, Justin, Rebekah and her family, Ben and his family, and Sherrie and I were the only ones there. They were all familiar with each other. We never asked why we were chosen to assist. With how popular these channels are, I'm sure that a lot of supporters applied to assist. No one knew us, to our knowledge.   We still don't know.

We got a ton of work done. We talked, worked, and had fun. The icing on the cake was Ben cooked the most amazing burgers I have ever had. I found out this is what he does. I was hating a little on his grilling skills before then. He buys the best beef from a local farmer and serves it out of his food truck. He's an amazing cook and an amazing, infectious kind of person. I met Ben's parents, who were there visiting from California, and was able to see where he got it from.

Everyone worked hard to get the work done. I could not help but notice the personalities of everyone there. They were real, genuine people. When I looked

BOBBY AND SHERRIE BLACK

around and saw the kids, it told me everything about the parents. As known and successful as all these content creators are and as weird as I know some people can be, they chose us to help, and we were honored.

I learned so much about growing and homesteading just that day, including a great way to amend the soil. We applied humus compost to Bri's new growing space. Justin supplied huge bags of compost that were incorporated into the ground. The ground when we arrived was covered with a black weed barrier that killed all the vegetation in the area that was to become the garden.

This was so simple but something I had never thought of. I had to tell Justin how I felt about what he was doing. To my surprise, they started talking about us coming back, and we made two more trips up that summer. I felt with each trip everyone knew who we were as people and the normal little hesitations disappeared.

Bobby, Justin and Rebekah Rhodes and Sherrie

Our Grocery Store

Sherrie

We are now growing and raising our food, such as collards, broccoli, cabbage, long beans, okra, kohlrabi, corn, green beans, six types of peppers, jicama, many types of watermelon, many types of cantaloupe and melons, thirteen types of banana plants, plantains, only about ten varieties of dragon fruit now, miracle fruit, Surinam cherry, three varieties of passion fruit, two varieties of oranges, lemons, two varieties of grapefruit, mangoes, three types of pineapple, ice cream beans, three varieties of apples, peaches, and pears.

We are now growing all kinds of plants for medicinal purposes, such as blue butterfly pea, lemongrass, gotu kola, cacao, miracle berry, comfrey, soursop, moringa, corn, awapuhi, sage, mint, rosemary, thyme, lavender, roses, wild lettuce, cayenne pepper, holy basil, feverfew, echinacea, chamomile, stevia, lemon balm, marigolds, dill, beebalm, hyssop, parsley, tarragon, milkweed, toothache plant, calendula, and other herbs.

We make all our personal hygiene products, such as bar soaps, liquid soaps, shampoo, conditioner, deodorant, body butter/creams, toothpaste, mouthwash, hair oils, herbal bath salt blends, and more.

We make household cleaners, such as pine cleaner, tub/tile cleaner, oven cleaner, multipurpose cleaner, toilet bombs, glass/window cleaner, laundry soap, dish liquid, and a host of other items.

We cook all kinds of foods from scratch, such as bread, kombucha, cakes, cookies, cinnamon rolls, candy bars, chocolate, ice cream, cheese, soda, juice, apple cider vinegar, peanut butter, jams/jellies, butter, plant-based meats, condiments, cereal, seasonings, just to name a few.

Grow What You Eat

Bobby

We've learned so many things since getting started on this journey, and learning how to grow our food more efficiently is one of the most important. So what would I tell someone new to growing and wanting to get on their journey of growing their garden? Well, there is no set of instructions that will fit everyone. Each person's journey will be unique. What I'm going to say will be specific to me.

First, everyone will grow in different circumstances. If you own property, you'll want to pick a location on your property that will be the most suitable for your garden. This location will differ for everyone. Avoid heavily shaded areas. Most plants that you grow will need sunlight to get the best growth. Not every plant will need all-day sunlight, so partly shaded will be fine. Keep in mind the sun doesn't shine equally across the globe. Florida sun and North Carolina sun are different.

Once you find that location, get a soil sample to make sure the location is suitable for growing. Send your soil sample to the appropriate place to get tested. Usually, the package will have instructions on where you can send it to be tested. You can also build or buy raised bed gardens. If you go this route, you will have to purchase high-quality soil for all your beds.

Purchasing good soil that has plenty of organic material added will save you that step. While purchased soil is a safer bet it will be a good idea to get it checked also. This can get costly but rewarding. Building a raised bed can prevent pests that live in your soil from feasting on the roots of your hard-earned crop. For those of you that don't have property and will be growing on balconies or porches, container gardens are what you need. The same soil that you put in the raised beds would work for the container garden pots. This soil will have nutrients and hold a lot of moisture.

So, you have a location and you have the soil right for growing. If you are new at this, don't bite off more than you can chew. You want to grow your food and you want to be successful at doing so. So, pick a crop to grow. The crop that you decide to grow will depend on when you are starting your growing and where you are located.

Where we live, we can start growing pretty early in the year. We can start seeds in January or February inside with the use of lights and heating pads. We usually can put our starts or plants in the ground in mid to late February or March, depending on the weather. We will get our best results on things like tomatoes and beans the earlier we put them in the ground. We typically have fewer pest issues during this time and the weather won't be so hot.

But this is much too early in other parts of the country. You'll have to research the best planting times for your region, as well as what plants are seasonal for when you want to start. We found this out through trial and error. Having a community to keep you motivated and a community close by that can give you information like this can be beneficial.

An online community will ask and answer questions that you may need to know, such as what kind of bug is this? Is this bug friend or foe? How do I get rid

ADVENTURES IN HOMESTEADING

of this bug? What is an organic way to get rid of this bug? What's wrong with my plant? And so on. They will assure you when you fail at something and feel down that it's normal for learning. The enthusiasm from your online community will keep you motivated to keep on pushing, and they'll give better advice on what to grow in your area and when.

Pick a crop that you and your family would like to eat that also grows well in your area and find out the right time to grow it. I suggest, if possible, starting with starts from local trusted sources. You'll be new at this, so avoid as many opportunities to fail as possible. Don't grow any more than two different things at first.

This will prevent you from having to concentrate on too many things at once. Different crops require different feeding, watering, sunlight, and pruning to grow properly. This can differ for every climate and different locations in the same climate. When looking at directions on how to grow, you should know these are guidelines. When you grow just a few crops at a time, you get to learn how to recognize issues and what to do to fix them.

We say that anyone who is thinking of growing their food should start now. The future is uncertain, and you never know what a new day will bring. After all, we never saw the pandemic coming, where the stores were closed and there were limited resources available. What is going to be the next issue that's going to take us by surprise? When will it be? How bad is it going to be? Did I prepare my kids and or family for this if I'm not around? On our homestead, we wanted to supply meat, partly because we wanted to know what we were eating and how it was treated.

We knew cows and some of the bigger animals

BOBBY AND SHERRIE BLACK

were going to be off-limits due to the size of our homestead, even though we raised pigs at one time. So we looked at what was doable for us and settled on rabbits, chickens, and quail. We didn't realize the other uses of these animals could be valuable to us on the homestead. Chicken manure for one is a great fertilizer for the garden that stops you from having to use chemical fertilizers. This will also save you money. Chicken manure's NPK (Nitrogen Phosphorus Potassium) value is, respectively, in percentages, around 1-1-0.5. Chicken manure is not safe to just use on your garden straight out of the coop and needs to be composted to be made safe for the garden.

We found that my banana plants and my dragon fruit don't mind hot chicken poop. That is a great resource that we already have because of having chickens. We just need to know how to use it. For comparison, quail manure's NPK is about 4-4-2, and it's also not good to use unless it has been composted. Then we come to our favorite and our main source of meat, our rabbits, and their manure's NPK value is about 2-1-1.

The amazing thing about rabbit poop is that it is cold, unlike the chicken and the quail manure, which means rabbit poop can be added straight to your garden and doesn't need to compost. Rabbit poop is four times as valuable as cow and horse poop and twice as nutritious as chicken. It is an amazing fertilizer that will help you grow an amazing garden without the worry of damaging it.

These are just a few of the things you learn when entering a community of homesteaders and growers. You learn how to properly compost. I'm not ashamed to say I have not mastered this yet, but that's okay, because I'm always learning and growing.

ADVENTURES IN HOMESTEADING

There are so many things that homesteading on any scale will help you with. Health for you and your family should be very important. If there is something you can do to make sure your family is as healthy as they can be, you should do it. Raising your meat is one of those things you should do if you can. If you can't, find someone you trust locally that sells the meat that they raise. I don't like speaking about large-scale farmers.

I respect what they do and know that it is necessary to feed the world. Unfortunately, some of their methods affect the food they produce, such as the conditions in which some of the animals are kept and the kind of chemicals, like steroids and antibiotics, that these animals are injected with. We are what we eat.

What are these chemicals doing to you and your family in the long term? Anyone who raises meat will tell you that the meat that they typically raise doesn't have the color of some of these store-bought meats. Well, the store-bought meats are treated with chemicals to allow the meat to look fresh.

Things like carbon monoxide keep the meat red even as it goes bad. What does that do to you? A high percentage of the chicken bought in stores was tested and contained dangerous bacteria. Let's not even talk about the conditions in which they are raised. It's just awful. The world must eat, and the more we need, the less the quality will get.

Do we blame the people who are producing this, or do we blame ourselves for not lessening the demand by trying to grow our own? Some say go vegan, but I like meat and don't want to give it up, so my answer is to rely less on what is coming out of the store. We use our little land to help us avoid the store as much as possible.

BOBBY AND SHERRIE BLACK

Let's not forget to look at the issues of not growing your vegetables. This is not much different than raising your meat. Many things are done to your vegetables, so they grow big and strong and do not have issues with pests. Again, what are these chemicals doing to your and your family's health in the long term? Some say, "I eat organic." What is organic? How do you know it's organic? Is it because that is what the label says, or do you know the people who grew it?

You may just need to ask yourself, does the organic gardener care more about what he grows being organic or his crop selling so he can make money for his family? The only way you'll know for sure is if you grow it yourself. We try and use our little bit of land to do just that, lessening the risk of consuming something we want to avoid.

We want to be clear that we live in this world, and we consume things like everyone does. We just try and do it less, so we don't add to any health problems that come about. We are not the kind of people who say that if you are not doing what I'm doing, believing what I'm believing, saying what I'm saying, then you are wrong, bad, stupid, or any of that.

ADVENTURES IN HOMESTEADING

Our conversations are for those who are like-minded. We very much are not going to do anything just because someone said we should. If someone mentions something to us, we research it, and if it applies to us, then we do it. There are so many out there that feel it's my way or the highway, but this is not our way. We know there are many ways to do the same thing.

Not accepting small differences in others stops you from evolving and growing. Our community is diverse, and we want it to be. Life has taught us that the things we deem important are much the same as those of any other person.

So Fresh and So Clean

Sherrie

We do "grow" and make our hygiene products! We also sell them. I'm so fascinated by soap making so I kept watching videos and learning.

One day I told Bobby we should make soap! We checked out the benefits of the things we grew and decided yes, this would be interesting. We came up with Blue Butterfly Honey Turmeric Soap and Honey Turmeric Banana Soap. Worked for weeks to formula these products perfectly. We wanted gentle but effective products, and we nailed it.

We can't keep soap on the shelves. We created a mild scent that the most sensitive can handle and still enjoy the amazing benefits. Have you ever checked the benefits of some of the plants you grow? Let's go over a few of our plants and their benefits to the skin.

We raise bees and harvest honey. A few benefits of honey for our skin are it deeply moisturizes, it's an antiseptic, contains antioxidants, and its anti-wrinkle. The bees draw moisture from the air adding moisture to the honey that can hydrate the skin giving us that beautiful, fresh, youthful, and glowing.

Turmeric can brighten dark spots, reduce blemishes, heal wounds, fade scars, moisturize, and provide that glow we all

ADVENTURES IN HOMESTEADING

love, and let's not forget the anti-aging priorities.

Banana Blossoms hold a rich source of antioxidants that protect the skin from free radicals and prevent fine lines from forming, it's enhanced with potassium, vitamin E, and vitamin C which lightens pigmentation.

Blue Butterfly Pea is known to stop the harmful effects of free radicals that damage and degrade skin.

Shampoo Ginger AKA Awapaui replenishes your skin's natural moisture, making it soft. Plus it's known to moisturize the skin and make you look younger so it also works as an anti-aging ingredient.

Now do you understand why we use these amazing products in our soap formulas? Combined it's a powerhouse for beautiful skin. I feel spoiled that we can grow these plants and use them in our hygiene products.

We also make Castile Soap! This is our liquid soap that can be used to wash your hair, wash your body, wash your clothes, wash your dishes, and clean your house. It's a gentle plant-based soap. It's like liquid gold. We don't scent this one because you may want to purchase this for your baby, or maybe to wash your car so if it's unscented you can choose to add essential oil of your choice or keep it unscented.

I should probably put a warning with soap making! It's a very enjoyable hobby and highly addictive. It warms my heart to know how to make such amazing products and off them to the world to help us get away from so many harmful chemicals.

home pantry

Oreo Cookies

1 cup unsalted butter, softened
1 cup white sugar
Cream together then add:
2 teaspoons salt
1 large egg
A splash of vanilla
2 cups all-purpose flour
1 ¼ cups dark cocoa powder
½ teaspoon baking soda

Instructions:

Mix wet ingredients and dry ingredients in two separate bowls, being sure to combine both well.

Then slowly combine dry ingredients with wet while blending. The mixture will be thick, sticky, and a little crumbly.

In 2 separate pieces, flatten & wrap in saran wrap, refrigerate for 1 hour

Roll out thin, cut & place on cookie sheet lined with parchment paper. Bake 350° for 8 minutes or so, depending on thickness. NOTE: Be careful—they can burn fast & it's hard to notice in the oven.

Cream Filling

1 stick butter, softened
2 cups powdered sugar

BOBBY AND SHERRIE BLACK

1 teaspoon vanilla

Pinch of salt

1/4 shortening, softened

Mix & play with measurements to get consistency for yourself

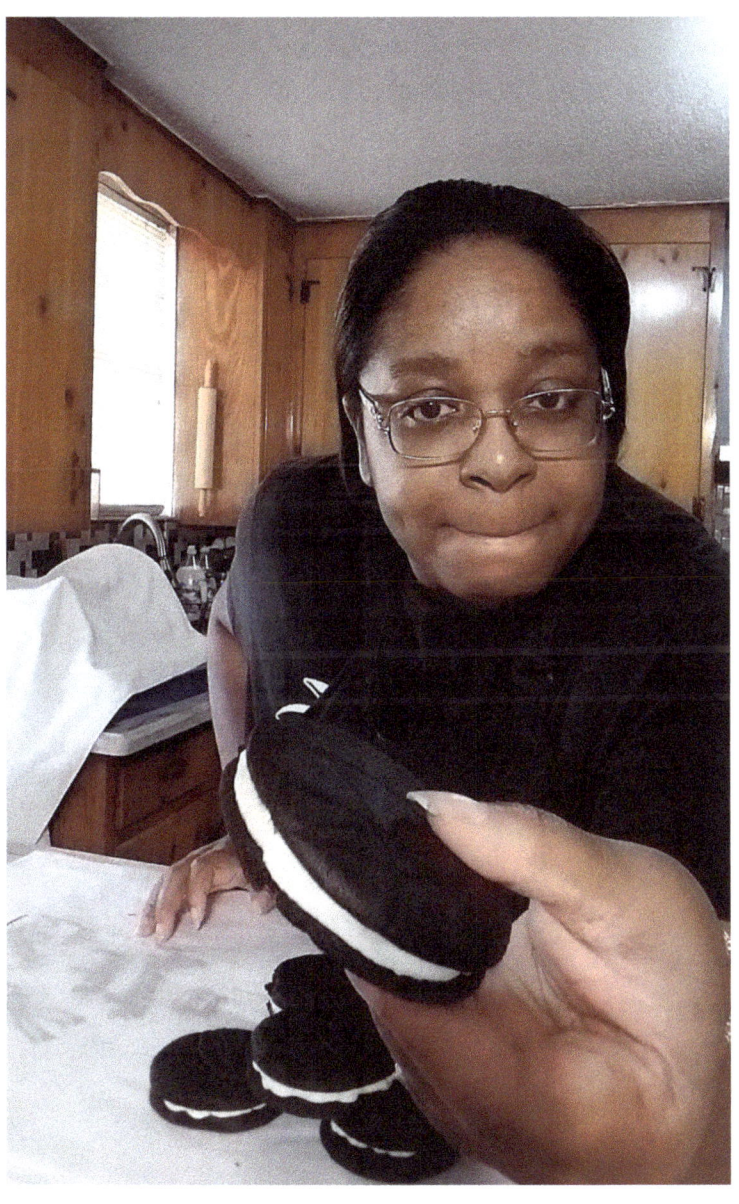

Cinnamon Roll

Mix:

1 cup warm water

1 tablespoon active yeast

1 tablespoon honey or sugar

Let bloom

Then add:

3 cups all-purpose flour

Mix well & let rise (double in size)

Then mix together enrichment ingredients:

1 cup mashed potatoes (dry or cooked)

1/2 cup sugar or honey

Dash of salt

1/2 cup butter (melted)

3 eggs

Add flour mixture, enrichment ingredients & 6 cups of flour together, knead until smooth, and the dough bounces back if you press into it. Grease with a little oil and let rise until doubles in size.

For a sticky bottom on the stove, I melt a stick of butter and 2 cups of brown sugar poured in the bottom of my pans, and top with nuts (optional).

When the dough is ready, roll it out and spread butter, brown sugar, and cinnamon all over. I add walnuts and pecans for Bobby.

I usually roll the entire dough into a log, or you can cut strips and roll one at a time. If rolled into a log, I use string to cut into size cinnamon rolls and place them in a pan. If cut into strips then rolled, I cut with pizza cutter.

I froze a few dozen! If we are cooking anything I let rise & bake 350° for about 25-30 mins.

Cream Cheese Drizzle:

Cream cheese, powdered sugar, and heavy whipping cream—mix to get the consistency you prefer

Vanilla Ice Cream

2 cups heavy whipping cream
1 cup milk
1/2 cup sugar/sweetener
1 tablespoon vanilla
Dash of salt

You can add any flavor, such as candy, fruit, or syrup.
www.youtube.com/watch?v=Wu6-ubXKWfE

Check out our new website.
www.BlacksTropical.com

Sandwich Bread

3/4 cup warm whey, water, or milk
1 tablespoon instant yeast
1 tablespoon honey or sugar
Let sit 5 minutes then mix:
Pinch of salt
1 egg
4-6 tablespoons melted butter
3 cup all-purpose flour
1/2 cup mashed potatoes

Knead until dough bounces back, then let rise for one hour. Cut dough into pieces, your preference on size for buns or rolls, or roll into log for loaf. Place in a greased pan. Let rise 30 minutes. Butter top and sprinkle with sesame seeds/everything bagel. Bake until golden brown at 350° about 30 minutes.

To freeze, use a freezer bag to store dough in the freezer until ready to bake, then take out of freezer, place dough in greased loaf pan, cover & let thaw, rise & bake. Dough is good in freezer up to 3 months.

The Most Beautiful Way to Start & End The Day Is With A Grateful Heart! God Is Good

ADVENTURES IN HOMESTEADING

Peanut Butter Cup

1.5 cup graham crackers crushed

2 cups powdered sugar

12 tablespoons butter

18 ounces of peanut butter, either creamy or crunchy is fine

8 tablespoons butter divided

24 ounces semi-sweet chocolate chips; you may want to have a little extra on hand just in case

In a medium-sized bowl, combine graham cracker crumbs and powdered sugar until well blended. Set it aside.

In a microwave-safe measuring cup, heat 6 tablespoons of butter and peanut butter together in the microwave for 15-second increments until melted. Stir until combined and pour over the graham cracker mixture. Continue to stir until well combined. Set it aside.

Melt half of the chocolate chips with 2 tablespoons of butter in a microwave-safe bowl in the microwave for 15-second increments (stirring each time) until the chocolate is melted.

Place cupcake liners in cupcake tin (mini or regular depending on what size peanut butter cups you want to make—make about 36 mini cups or 18 full-sized cups).

Using a paintbrush (I used a flat brush about 1/2" wide), "paint" each cupcake liner with melted chocolate. Be sure to evenly coat the bottom and halfway up each cup (if using the "mini" cupcake

ADVENTURES IN HOMESTEADING

liners). If you are using full-size cupcake liners, paint about ⅓ of the way up each cup.

Tip: You don't need to paint the chocolate super thick, just enough so you don't see any of the liner underneath.

Place in the refrigerator to cool for about 5 minutes or until chocolate hardens.

Press about 1 teaspoon of graham cracker mixture in each mini cup (about 1 tablespoon for regular-sized cups) making sure the top is somewhat flat (it doesn't need to be perfect). Just make sure you don't go above where you "painted" the chocolate on the cup.

Add the remaining chocolate chips to the remaining melted chocolate in the microwave-safe bowl and melt together with the remaining 2 tablespoons of butter (stirring every 15 seconds).

Spoon a little chocolate over each peanut butter cup until the peanut butter filling is completely covered (you can lightly shake or tap the cupcake pan to even out the

chocolate over the filling). You want the top of the peanut butter cup to look completely flat.

Place in refrigerator or cool area until chocolate hardens, then serve!

Keep Your Eyes On The Prize & Not The Obstacles!
Good Morning SoilBrotha's & SoilSista's, Many Blessings To You All

BOBBY AND SHERRIE BLACK

We Love & Appreciate You

Honey Caramel

1 cup raw honey

¾ cup heavy whipping cream

1 ½ teaspoon vanilla

6 tablespoon butter

After honey and heavy whipping cream come to 240-250°, take off the heat and add butter and vanilla. Line pan with parchment paper or fill molds, set in freezer or refrigerator. Caramel will be soft or melt in the heat.

Chocolate

¼ cup cacao/cocoa powder

3 tablespoons honey

3 tablespoons coconut oil

Mix well until combined and enjoy!

Ginger Bug (For Homemade Soda)

3 tablespoons organic ginger (chopped)

3 tablespoons cane sugar

2 cups non-chlorinated water

Chop or grate ginger, add to jar.

Add sugar and water to jar and mix.

Cover with breathable cloth for 24 hours.

On day 2, add 1 tablespoon of chopped ginger and 1 tablespoon of sugar.

Cover with breathable cloth for another 24 hours.

On day 3-5, repeat steps 4 and 5.

On day 6, your bug is ready and can be refrigerated (feed bug weekly in refrigerator 1 tablespoon cane sugar, 1 tablespoon organic ginger)

Note: When the bug gets low, you can always add more non-chlorinated water, 1 tablespoon cane sugar & 1 tablespoon organic ginger.

Ginger Ale

½ gallon filtered water (chlorine kills the beneficial yeast and bacteria)

1½-2 cups sugar (organic white sugar or unrefined sugar will both work; you can add less sugar depending on your taste)

Cups of organic fresh ginger root, chopped.

1 cup ginger bug starter

⅓ cup lime juice or lemon juice (optional)

Place the water, sliced ginger root, and sugar in a pot and bring to a boil.

Reduce heat to low/med and simmer for 10 minutes, add ½ gallon of water (one gallon total) then set aside and allow the tea to cool to room temperature. Keep in mind that if the ginger tea is too hot when you add the ginger bug, it will kill the beneficial yeast and bacteria needed for fermentation.

Place a fine mesh strainer over a bowl and pour the mixture into it to separate the ginger from the liquid.

Add the ginger bug and lime/lemon juice (if using) to the liquid and mix with a wooden spoon.

Pour the liquid into flip-top bottles (leaving one inch of head space) and ferment for 3-5 days. During the fermentation process you'll see bubbles, by day 5 lock bottle tops and "burp" the bottles by opening them every 1-2 days so the carbonation doesn't build up too much as you want (we do not burp). Some people have reported that their bottles exploded because they let them ferment for way too long without burping them. I've been making fermented drinks for a while and have never had that happen, but I forgot about them for a few days and ended up with an extra fizzy bottle that poured out like champagne. I open my bottles on our deck now just in case.

The ginger ale is ready when it's fizzy and not overly sweet. If you check it after a week and it's still too sweet, you can let it ferment for a bit longer so that the good guys can consume more of the sugar and turn it into probiotics.

Notes:

This recipe will fill approximately 8-9 of the swing-top bottles pictured (in the video) or 4-5 of these larger swing-top bottles.

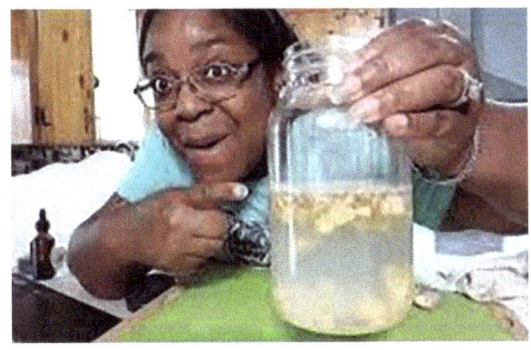

If you don't have swing-top bottles and prefer to use what you have on hand, opt for a jar that gets a good seal when the lid is on. Fido jars are a good option, but mason jars will work, too. A ginger ale that is fermented in jars won't be very fizzy (if at all), but it will still be probiotic-rich and yummy.

Shampoo

½ cup castile soap
1½ cup coconut milk
½ cup distilled water
1 teaspoon apple cider vinegar
1 teaspoon coconut oil
1 teaspoon Vitamin E oil
1 teaspoon almond oil
2 tablespoons fenugreek powder

Mix well

Keep refrigerated for up to a month or freeze for up to 6 months.

Hair Conditioner

⅓ cup dried banana blossom
4 tablespoons coconut oil
2 tablespoons honey
Mix until well combined.

Apple Cider Vinegar (1 Gallon)

Apple/fruit scraps (sliced fruit if you choose)

1 cup cane sugar

Fill the jar with non-chlorinated water.

Mix well and cover with cheesecloth/breathable cloth/rubber bands set in a dark place.

Mix daily for 1-2 weeks.

After a month has passed, you can give your vinegar a taste test. If it tastes vinegary enough to your liking, move on to the next step. If not, allow it to ferment longer. When doubtful, you can check the pH of your apple cider with these simple pH test strips! Finished apple cider vinegar should have a pH in the range of 2-3.

We usually strain around 6-8 weeks.

Store in a temperate, dark location for at least one month, or longer. The bacteria will keep working to convert more and more of the sugar or alcohol to acetic acid, creating vinegar. The rate at which your partially fermented apple cider turns into full-blown vinegar will vary, depending on the storage conditions and the apples used. Our apple cider vinegar is usually ready in about 2-3 before we bottle, cap, and use it. Your time may differ.

Motivation Is What Get You Started, Habit Is What Keep You Going! You Got This SoilFamily, You Are Strong, You Are Amazing, You Are Loved

Laundry Soap (1 Gallon)

1 cup borax (freshens, deodorizes, and lifts dirt and stains)

1 cup washing soda (freshens and deodorizes)

1 cup liquid castile soap or Sal Suds (the main cleaning agent, lifts dirt and cleans)

50 drops essential oil (optional for a scent)

15 cups water (hot/warm)

For top-loading washer machines, use 1/8-1/4 cup.

For a front loader, use 1-2 tablespoons.

Every Morning Is An Opportunity To Create A Masterpiece Called Life!

Dish Detergent

⅓ cup washing soda

1 cup castile soap

1 cup hot water

15-20 drops essential oil (optional)

Mix until well combined

Ability Is What You're Capable Of Doing, Motivation Determines What You Do & Attitude Determines How Well You Do It!

Fire Cider

1 whole onion
1 whole lemon
1 whole orange
Horseradish
Cayenne pepper (dried or powder)
Garlic
Turmeric
Black pepper
Rosemary
Ginger
⅓ cup dried elderberry
½ gallon apple cider vinegar

Amounts are not set in stone, adjust to suit you and your family. Chop ingredients, place in half-gallon jar, pour apple cider vinegar, place top on, shake and sit in the dark, cool space for 6-8 weeks. Drain liquid, add at least one cup of raw honey, mix well, and enjoy.

We take 1 tablespoon of fire cider daily as a preventative. If sick, take 3 tablespoons daily until well.

Elderberry Syrup

Ingredients:
Dried elderberry,
ginger,
cinnamon,
vanilla,
dandelion root,
cayenne pepper,
black pepper,
honey,
cloves

Instructions:

Using A Cloth Bag/Tea Bag, place your
2 tablespoons of elderberry
1 tablespoon of ginger
1-2 cinnamon sticks
1 vanilla bean or 2 tablespoons of vanilla
1/2 teaspoon black pepper
1/2 teaspoon of cayenne pepper
1 teaspoon dandelion root
1/2 teaspoon cloves
Amounts do not have to be exact!

Bring 4 cups of water to a boil and place the bag in. Turn the heat down and let simmer for about 20 mins. Allow the tea blend to cool a little then squeeze

ADVENTURES IN HOMESTEADING

the bag to get liquid out and remove the bag. Add 1-2 cups of raw honey and mix well. Be sure to label your jar and refrigerate it after it's cool enough. Take daily to keep the Dr away, especially during cold and flu season. Compost herbs and tea bags.

Your Body Holds Deep Wisdom! Trust In It, Learn From It, Nourish It & Watch Your Life Transform!

Focus Tea Blend

Ingredients: Gotu Kola, Rosemary, Thyme

Using a Tea Bag, place 1/2 teaspoon of Gotu Kola, rosemary, and thyme inside the bag tightly securing the bag. Bring 16 oz of water to a boil and add the tea bag. Turn down the heat to a light simmer for 10 mins then turn the heat off. Allow the tea to steep for about 10 minutes before squeezing liquids out of the tea bag and removing it! Tea bags can go in compost. Add honey if desired and enjoy.

This beautiful blend has many benefits for brain health. Help with alertness, focus and so much more.

Your Life Is Full Of Purpose! Don't Let Nobody Steal Your Joy

Kombucha

First, let's talk about the SCOBY! WHAT'S A SCOBY?? Scoby is a Symbiotic Culture Of Bacteria and Yeast is a gelatin-like substance that forms on top of the brew, The Scoby is the "mother" that kickstarts each batch while also protecting the kombucha from contaminants. No honey: Honey can contain botulism.

You'll Need:

16 cups of non-chlorinated water

1 cup white sugar

4-6 bags of black tea

1 cup unpasteurized, unflavored store-bought kombucha

A large glass or ceramic container (at least a gallon vestal)

Breathable cloth (coffee filters, paper towels, napkins, cheesecloth)

Rubber bands

Large pot for boiling water

Instructions:

Bring 8 cups of non-chlorinated water to a boil, remove from heat, and dissolve sugar into it, add the tea bags and allow them to steep for at least 20 minutes then remove the tea bags and allow the tea to cool completely.

Add tea to the glass jar and the remaining 8 cups of cold water, which will bring the mixture to room temperature faster.

Then pour store-bought kombucha in, making sure to mix the entire mixture well. These are great for kickstarting The Fermentation!

Cover with a few layers of tightly breathable cloth to keep out bugs and debris, securing with a rubber band.

Set in a dark, still, and room temperature around 70-75 degrees for about 2 weeks or so until a ¼ inch or so SCOBY has formed.

CONGRATULATIONS YOU HAVE A SCOBY!!

He is ok in this jar keep covered with breathable cloth!! Now let's make some kombucha!

Instructions For Making Kombucha:

16 cups nonchlorinated water

1 cup white sugar

4-6 bags of black or green tea

1 cup unflavored kombucha from Scoby Hotel

1 SCOBY

This process is just like growing your Scoby.

Bring 8 cups of non-chlorinated water to a boil, remove from heat, and dissolve sugar into it, add the tea bags and allow them to steep for at least 20 minutes then remove the tea bags and allow the tea to cool completely.

Add tea to a glass jar and the remaining 8 cups of cold water, which will bring the mixture to room

ADVENTURES IN HOMESTEADING

temperature faster.

(Don't be impatient here – hot water will kill your SCOBY).

Once the tea has cooled add your Scoby and brine from Scoby Hotel to your room-temperature sweet tea cover with breathable cloth and secure with rubber bands.

Ferment: Set the jar somewhere dark, still, and at room temperature 70-75° for 7 to 10 days. Begin tasting at about 7 days. It should be mildly sweet and slightly vinegary. The warmer the air temperature, the faster the kombucha will ferment. The longer the tea ferments, the more sugar molecules will be eaten up, the less sweet it will be.

The final and most fun step in the homemade kombucha-making process! The second fermentation is where the real magic happens, flavoring and carbonating your kombucha.

Sweetener (fruit, honey, or sugar). While there are many flavor combinations, we generally work with a ratio of 1 cup of kombucha to:

1 to 2 Tbsp mashed or freeze-dried fruit

1 to 2 tsp honey or sugar

You just need flip-top glass bottles for the second fermentation. These bottles are meant for fermentation and have an airtight seal, which will prevent carbonation from escaping. If you don't have these, canning jars will do an alright job, even though they aren't airtight.

*Bottle: Funnel kombucha into bottles, leaving about 1 1/2 inches or so at the top.

*Sweeten: Add your chosen sweetener and seal tightly.

*Let ferment somewhere dark and at room temperature for 2-3 days

*Sat in the refrigerator to chill and slow down the carbonation process. If desired, strain out the fruit before serving.

Enjoy Your Healthy Probiotic Drinks

Lemon Oreo Pound Cake

Preheat Oven to 325

INGREDIENTS:

3 Cups Swans Down Cake Flour

3 Cups Sugar

3 Sticks Unsalted Butter (Room Temperature - let sit overnight if possible)

6 Large Eggs (Room Temperature)

8 oz. Cream Cheese (Room Temperature - let sit overnight if possible)

1 3.4 oz Box Lemon Jello Pudding & pie filling

15-20 Lemon Oreo Cookies (Crushed or pulsed using a Vitamix or food processor) 1/2 Tsp Baking Powder

1/4 Tsp Salt

1/2 Tsp pure vanilla extract

1 to 2 Tbsp. Pure Lemon Extract (depends on YOUR personal taste)

DIRECTIONS:

1) Sift cake flour, pudding mix, and baking powder. Set aside

2) In a large mixing bowl, stir butter, cream cheese, salt, and extracts just until combined (about half a minute). You can do this by hand OR use a mixer on LOW speed.

3) Turn the mixer up and begin adding the sugar. After all of the sugar has been added, mix on high for approx. 10 to 12 minutes until the batter is creamy and fluffy.

4) Add eggs ONE at a time; mixing JUST UNTIL the yolk is combined. DONT OVERMIX

3) Turn the mixer down and add flour 1 cup at a time. DO NOT OVERMIX. You simply want to incorporate the flour.

4) Stir in cookies by hand OR mixer. DON'T OVERMIX! (optional)

5) Spray bundt cake pan, fill, and place in the oven for 1 hour and 15 minutes. (Time may vary depending on your cake pan and/or oven.)

6) Cool for 15 minutes BEFORE you release the cake from the pan. Cool another hour before glazing OR frosting.

Cake Glaze - Ingredients

2 Cups of Powdered Sugar

1 oz of Cream Cheese (room temperature) 2-3 teaspoon milk

A teaspoon of lemon extract

Mix the cream cheese until it is smooth. Add the powdered sugar and lemon extract. Mix. Add the milk a LITTLE at a time just to give the glaze a good consistency. I only use a little because I like my glaze thick.

Crumble Oreo Cookies To Garnish Top & Enjoy

Always Know That You Are Unique, You Are Important, You Are Amazing, You Are Special, You Are Loved!

Natural Immune Booster

This Recipe Can Be So Beneficial To You & Your Love Ones. We Always Say Do Your Research, This Is NOT MEDICAL ADVICE! Ingredients, Take Away/Add What You Like.

Ginger

Garlic

Onion

Cinnamon

Cayenne

Turmeric

Black Pepper

Lemon(With rind)

Apple Cider Vinegar

Honey

Keep Refrigerated Up To 6 Months. Take Daily As A Preventative. Stay Well

The Best Love Is The Kind That Awakens The Soul & Makes Us Reach For More! The Kind That Plants A Fire In Our Hearts & Beings Peace To Our Hearts!

Watermelon Jam

Ingredients:

1 cup watermelon puree(fruit of choice)

3/4+ cup sugar(depending on how sweet you like)

4 tablespoon pectin

2 tablespoons lemon juice

Puree Watermelon (can use rinds also just adjust your sugar)

Add to medium heat sugar(your preference on how sweet) and let the sugar dissolve

Add pectin mix well so pectin does not thicken in one spot

Add lemon juice and let simmer for about 10 minutes

Put the mixture in a jar in a water bath can or sit to cool before refrigerating

It is ok on the counter after canning but should be refrigerated after opening

Please follow the Ball Canning Guide for proper canning practices

Frosty

Ingredients:

1/2 c flavored milk (chocolate, banana, strawberry)
14 oz. Sweetened condensed milk
8 oz cool whip
1 tspVanilla Extract

Mix ingredients until well combined

Place in a freezer-safe container and place the container in the freezer

Check every hour mixing mixture until solid

Enjoy Your Frosty Without Leaving Home

Homemade Cheese

Ingredients:

1/2 Gallon Whole Milk

7 tablespoons distilled White Vinegar

Instructions:

Pour whole milk into a large Dutch oven

REMEMBER ultra-pasteurized or homogenized. Ultra-pasteurized and homogenized won't Work Properly

Set the heat in between medium-low and medium. We don't want to heat the milk too quickly, heating it to 115°

Once the milk gets to that temp, turn off the heat, and add in the vinegar. Stir in the vinegar for about 30 seconds so it evenly mixes in with the milk. It will start to curdle almost immediately.

Add a lid to the pot and allow it to sit undisturbed for about 10 minutes.

Take a spoon or spatula and bring the curd to the side against the pot. Make sure to gather all the whey

BOBBY AND SHERRIE BLACK

As you gather it, push it up against the wall of the pot. Then remove it from the whey and place it in a bowl. Time to remove as much whey as possible. Using very clean hands or gloves, grab the curd cup it in your hands and gently squeeze. You can use a cheesecloth if you choose. This will press out the whey.

Keep repeating this process several times, using a bit more force each time. You can also place it in the bowl and knead it a bit pouring the whey into a separate container or bowl.

Once you have as much whey as possible pressed out of the cheese, place the cheese in a microwave-safe bowl and hear your cheese for 30 seconds at a time bringing cheese up to 160°

You can also reheat your whey and submerge the cheese for 20 seconds to heat the cheese up

Once your cheese reaches 160° immerse the cheese ball into a bowl of cool water for about 10 minutes then you can season your cheese and place it in the refrigerator to chill.

Root Beer Soda

First:

1/2 gallon filtered water

1 tablespoon ginger

2 Cinnamon sticks

Boil for About 15-20 Mins Then Take Off the Heat

Add:

1 tbsp. Dandelion root

2 tbsp. sarsaparilla root

1-star anise

1 tbsp. licorice root

2 tbsp. sassafras root

Let Steep for About 20 Mins Then Straight Herbs

Add:

1/2 cup molasses - Mild and unsulphured (not blackstrap)

1 cup granulated sugar

1 cup ginger bug

After RootBeer syrup has cooled down add 1 cup ginger bug for fermenting

For Fermenting we let ours sit uncapped for about 2-3 days then we cap to build carbonation for another day or two before refrigerating.

BOBBY AND SHERRIE BLACK

There are other herbs you can use to make RootBeer, these are the herbs of our choice. Please Do Your Research In Each Herb.

Thank you for watching, Share & Enjoy

Bath Cleaning Bombs

Silicone molds

4 ounces baking soda

2 ounces cornstarch

2 ounces citric acid

2 ounces Epsom salt

1 to 1 1/2 teaspoons water

5-10 drops essential oil of your choice

Wearing gloves, mix all ingredients in a bowl until well combined

Fill molds with mixture being sure to pack tightly

Allow to sit 24 hours to harden

Remove bombs from molds and place them in an airtight jar for storage

Sometimes The Strongest Thing You Can Do Is Smile & Carry On! Today Is Going To Be A Fabulous Day

Bobby and Sherrie Black's

What's up Soil Brothas & Soil Sistas? We are Bobby & Sherrie Black, founders of Black's Tropical Homestead and SoilFamily Expo Inc 501(c)3. We own our one-acre suburban homestead, located in Savannah, Georgia, where we grow and raise our groceries. We have a special love for growing tropical fruit, vegetables, and herbs, making homemade items, and sharing our adventures with the world. We also love to travel, fish, and listen to live bands.

Our mission is to help fight food insecurities by encouraging and helping our soil family all over the world to homestead/garden/farm where they are, no matter the size of their property. In this world of uncertainty, we hope you join us on this amazing adventure to live a healthier, peaceful life. We love and appreciate you.

www.ingramcontent.com/pod-product-compliance
Lightning Source LLC
Chambersburg PA
CBHW050735010526
44107CB00010B/860